STAR WAVE

The mind of man is capable of anything because everything is in it. All the past, as well as all the future.

—*Joseph Conrad*

Star wave is a wave from the future. In quantum physics anything physical is represented by a wave of probability called the quantum wave function. To determine these probabilities this wave must be multiplied by the star wave, the complex-conjugate, time-reversed, mirror-imaged wave. In brief, the star wave makes dreams come true.

The frontispieces beginning the chapters are frames from the motion picture of a star wave of two correlated particles. Through such correlations, the mind is at work.

STAR WAVE

Ψ*Ψ

Mind, Consciousness, and Quantum Physics

FRED ALAN WOLF

Macmillan Publishing Company

New York

ALSO BY THE AUTHOR:

Taking the Quantum Leap
Space-Time and Beyond (with Bob Toben)

This book was written by the author and typeset by the Publisher using "Macmillan Writer," a software program developed under the direction of Macmillan Graphic Information Services. This project was carried out under the supervision of Casey Lee and Bob Keefe.

The following drawings were reproduced with the permission of the artists: "Three Theories" by Richard Cline (page 33); "Enter the Variable" by Peter Mueller (page 98).

The drawing "Parallel Universes" by Roz Chast (page 228) was reproduced with the permission of the artist and *The New Yorker* magazine.

The poems by Carl Sandburg, "Number Man (for the ghost of Johann Sebastian Bach)" (page 84) and "Wilderness" (from *Cornhuskers*) (pages 266-67), were reprinted courtesy of Harcourt Brace Jovanovich.

MACMILLAN PUBLISHING COMPANY
866 Third Avenue, New York, N.Y. 10022
Collier Macmillan Canada, Inc.

Library of Congress Cataloging in Publication Data

Wolf, Fred Alan.
Star wave.

Bibliography: p.
Includes index.
1. Quantum theory. 2. Physics—Philosophy.
3. Mind. 4. Humanistic psychology. I. Title.
QC174.13.W65 1984   530.1′2   84-17101
ISBN 0-02-630860-6

Macmillan books are available at special discounts for bulk purchases for sales promotions, premiums, fund-raising, or educational use. Special editions or book excerpts can also be created to specification. For details, contact:

Special Sales Director
Macmillan Publishing Company
866 Third Avenue
New York, New York 10022

10 9 8 7 6 5 4 3 2 1

Printed in the United States of America

CONTENTS

PART FOUR:
NEW FRONTIER OF THE MIND

PREFACE

Star Wave is a message of hope. It is my attempt to explain as best as I can the real reason we human beings are here in the first place, as conscious, spiritual creatures. I believe that quantum physics, the most powerful and rigorous science devised to date, will provide a basis for the formation of a new psychology—a true humanistic psychology. The tools of quantum physics, invented by the human mind, are turned around to investigate the mind itself. This is, as far as I can determine, the first attempt to bring psychology into the light shed by the discoveries of quantum physics.

Star Wave contains new ideas—ideas that have never found their way into print before. They may appear bizarre at first. These new ideas include the notions that energy and feelings, intuition and momentum, spatial location and body sensation, and time and thought are equivalent pairs. I ask only that the reader give these ideas some thought.

This will not be easy or light reading. In spite of its difficulties, the nonscientist will garner rewards. I have attempted to mark the most difficult passages, those that have some mathematical underpinning, by placing a gray line alongside the left margin. The reader may wish to skip over the gray-lined material and return to it at a later time.

I hope to demonstrate to the reader:

- how the future is more important than the past in deciding the present,
- that the future already exists while the past is continually being re-created,
- that time, as we presently understand it, is an illusion,
- how artificial and human intelligence differ,
- what creativity consists of in terms of quantum physics,
- how evolution is a consequence of the future and not of the past,
- how parallel worlds appear to us as thoughts and psychic experiences,
- how the ideas of determinism and causality arose and how and why they are incorrect,

- how nature creates repeated patterns that are taken as proofs of causality,
- how quantum physics provides a basis for a new religion and an understanding of the human spirit,
- why disagreement between peoples, states, and nations are bound to occur, and how they can be rectified.

In brief, it is my hope that this book will start a revolution of human thought. It is not too late. It is about time.

ACKNOWLEDGMENTS

It is difficult to acknowledge all of the people who have influenced and inspired me to write this book, which represents nearly fifteen years of my thinking. During this time I have changed and, consequently, so have my ideas. *Star Wave* contains my latest visions and concepts.

To the following persons throughout history I owe my gratitude: Judith Wolf, my wife and partner, who had the courage to see this through and has continued to support, love, and nurture me throughout the difficult years of my career change.

Carol Ann Dryer and Ivan Dryer, my dear friends, who also nurtured, sustained, and comforted me through many of the difficult years.

Bob Toben, my longtime friend and past benefactor, who always offered a helping hand or word when it was needed.

John Archibald Wheeler, for his vision and for his naming of the new age of physics, The Physics of Meaning.

Sir Fred Hoyle, for realizing that the future decides the present.

Physicists Albert Einstein, Niels Bohr, Werner Heisenberg, Louis de Broglie, Paul Adrien Maurice Dirac, Eugene Wigner, Richard Feynman, and David Bohm, for their visionary contributions to quantum physics.

Ernst Heinrich Weber and Gustav Fechner, for their pioneering efforts in the beginnings of psychophysics.

Sigmund Freud, who may be out of favor at present, but was the first mind scientist.

Carl Jung, for starting a revolution by bringing God into psychology.

B. F. Skinner, for attempting to make psychology an objective science.

Da Free John and Carlo Suares, for providing spiritual wisdom and evidence that quantum physics is the physics of the human spirit.

Joe Libs, who lent much support by providing books, resources, and valuable insights into the world of publishing.

John Brockman, my literary agent, for being the best agent and for his continuing advice.

Charles Levine, my editor, for understanding this and for his gentle "blue pencil."

The staff at Macmillan Publishing Company, for their continued bravery in the publishing world.

To all visionaries of the future who guide these words now.

ix

MINI-GLOSSARY

Qwiff. A term coined by the author for the quantum wave function. It is a wave that contains the potential for anything physical to appear. It is abstract and unobservable, but when it "pops," the physical world manifests.

Quamp. A term coined by the author for the quantum probability amplitude. It is the stuff that qwiffs are made of. It is composed of two parts — a "bra" and a "ket." The ket is an initial question or condition. The bra is the final question or condition.

Bra. The leading "edge" or future aspect of a quamp. It is written $<B|$ and reads B-bra. It is the possible condition that a state B is realized in the future.

Ket. The trailing edge of a quamp. It is written as $|A>$ and reads ket-A. It is the initial state that A is realized now.

Bra-ket. Put a bra and a ket together and they spell a possibility. For example, if A stands for the state "I am hungry" and B for the state "I am full," then B-bra – ket-A, written $<B|A>$, is the question, "If I am hungry, then am I full?" In this case the answer is no, or zero. What is needed is an operator to "sandwich" itself between A and B.

Operator. Something that works or "operates" on a ket and can change it. For example, the operator E (the "eat" operator) operates on ket-A, $|A>$, making it into a ket-B, $|B>$, which is written:

$$E|A> = |B>$$

Or, after eating I am full. An E operator sandwiched between a B-bra and an A-ket makes for a full stomach and a probability of certainty, which is written:

$$<B|E|A> = <B|B> = 1$$

Parallel world or universe. An alternative to a qwiff "pop." Instead of a qwiff, which represents all possibilities, suddenly manifesting one possibility, according to the parallel world alternative, all possibilities are in actual existence simultaneously, as if in parallel worlds unbeknownst to each other.

Hilbert space. The abstract mathematical space of all possible bras and kets. The Hilbert space is the space of the mind. Just as ordinary space is spanned by directions called vectors, the mind space is spanned by bras and kets. These bras and kets form the "directions of our minds" and appear to us as ideas.

PART ZERO

$$\Psi * \Psi$$

AN INVITATION

An invitation is offered to cross a bridge as vast as the infinite space surrounding us and as small as the width of a neuron's membrane. This is a journey into unknown territory, a voyage into the mind. To guide us we shall use the lanterns of modern physics, and the guides will speak to us in the language of the new physics. Not only is the territory strange, awesome, and wonderful, but we shall learn to see it in the unfamiliar terms of the physics of today, quantum mechanics. To know the mind is to know the universe.

The Mind of a Physicist

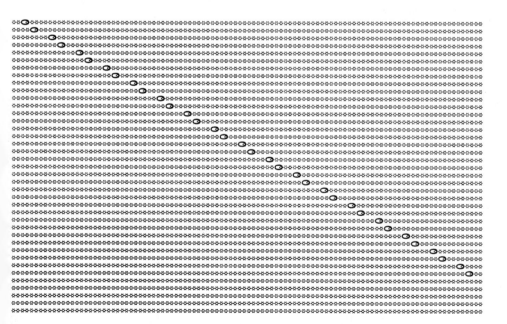

But our spurned senses reply, "Wretched intellect. You get your evidence from us, and you try to overthrow us? Your victory is your defeat."

Democritus

One Question and One Question Only

This book deals with a difficult and often controversial subject: the relationship of mind to matter. Inherent in this relationship lies a single mystery whose solution answers all other mind/body or mind/matter questions: How is it that I exist? This engenders a related question: How is it that anything exists? As a result of my years of study and research in physics, particularly quantum physics, I believe that there is a new answer. This realization led to my writing this book.

Laws of Physics and Laws of Mind

A natural question to ask is, Why physics? That is, why might physics supply the answer to the mystery of the mind? To answer this, let me sketch a brief history of this branch of science. Physics was invented as a practical way to live, indeed survive, in the universe. It provided us with methods for understanding our interactions in terms we now call matter and energy. Undoubtedly, the first application of physics involved simple machines, such as levers and rolling logs. The simple formula "force times distance is conserved" taught us how to use a lever. By placing a fulcrum toward the end of a lever, beneath a weight to be lifted, we realized that our small human force times the larger distance to the fulcrum was equal to the smaller lever distance times the heavier weight. Thus we experienced force acting on objects as a result of our own effort-force. From forces we felt inside of ourselves came the idea that force existed outside ourselves.

In the opposite manner psychology was created from observations of the outside world. From experiences we saw occurring outside of ourselves came the idea that these experiences existed inside our heads. Psychology was created from human observation of human and animal behavior.

From forces we observed motion, how and why things move. The observation of motion led to the realization of action-at-a-distance. Forces could act without contact and could change motion. This led to the idea that a field existed in space that acted on an object in space. This idea resulted in electromagnetism, radio and television. Physics in itself unifies and abstracts from all practice an essence or an ideal. This essence is called a theory and a theory in turn uses mathematics and number for absolute clarity. Number is the clearest human concept.

With number and human experience as a guide, thinking people began to inquire about themselves. Clearly there was felt to be a division between the personal, or subjective, inner world and the objective outer world. Through

4

physics and its basis in number, humans began to trust their intuitions and their abilities to understand the outside world. Physics actually had its origin in early Greek thought and derives from the word *physis,* meaning spirit, and provided relief from doubt and uncertainty about the human condition.

No wonder, then, that psychology has borrowed so much from it. All modelers in the field of psychology use, at least to some degree, concepts borrowed from the physics of their time. By providing fundamental insights about how nature works — through ideas like force, energy, conservation of energy, equilibrium, disequilibrium — the concepts of the physics of each period in history permeate and underpin all fields of thought, from chemistry to medicine and psychology. This has led me to consider how the latest discoveries in physics, the "new physics," may provide great insight into the workings of the "psychical apparatus," as Freud put it. I believe that these discoveries will appear quite revolutionary and also provide the unification of several schools of thinking, including the Freudian, the Jungian, the various humanistic psychological schools, and Buddhist and Hindu schools, not to mention Judaism and Christianity.

What is the new physics today? How does it contrast with the old? To understand this we need to look at a previously unsuspected and unexamined assumption about the nature of observation of anything physical.

The old physics contained within it a certain idealized concept about the nature of observation. This idea was that observation happened irrespective of the observer. The observer was merely a recording instrument — a mechanical device. He or she was entirely a victim of circumstances, recording and acting mechanically on receipt of experience. One never even questioned the possibility that the observer could play any other role than a passive one. Coupled with this view was the concept that a thing or an object was not understood unless it was analyzed. Analysis was performed as a cutting operation, a splitting apart of the thing to see how it "worked." Naturally, psychoanalysis was to follow in a similar way, with a detached observer analyzing the patient's psyche.

It is interesting that the word *certain* comes from the Indo-European root *skeri,* meaning to cut or take apart.

But then came the new physics. It came but was not welcomed because it showed that cutting led to new experiences that were paradoxical when compared with the previous experiences. For example, it was recognized that by attempting to locate accurately the position of an atomic particle, the particle was so disturbed that no one knew where it would be found later. By not cutting such a fine edge, by not attempting to "pin down" the particle's location specifically, the particle was not disturbed as much. Consequently and paradoxically, one actually gained information about the particle's future by knowing less about its past.

Thus analysis did not necessarily produce a simpler or greater understanding of phenomena. With the experimental tools of high-energy physics coupled with the concepts of the new physics, one discovered more and more paradoxical behavior as a result of analysis. By colliding two subatomic particles, two protons, at extremely high energies, sometimes as many as six protons (four protons and two antiprotons) would be created. It was like banging and squashing two apples together and finding other apples inside that couldn't have been there to begin with!

The realization that analysis did not lead to greater understanding made us suspect that somehow we were responsible for our observations. What we brought to bear on observation was creating our experience: If I saw red, I was projecting red. If I saw something move, I had to be projecting momentum from myself in some way. How that is done is debatable, but few physicists would doubt that observation is involved in doing it in some manner. Indeed, in quantum mechanics we have created a new language to account for this kind of projection.

My study of quantum physics made me realize that it is a psychological science as well as a physical one. This realization followed from the fact that the observer had a dramatic effect, as a result of choosing what to look for (the principle of complementarity), on the results of his or her observations.

Following along this track, all so-called outside physical experiences must be accompanied by a mental experience if they are to be known at all. If we are to regard this seriously, we must conclude that there is a linking of events in the mind with the simultaneously occurring events in space and time. For example, the mind could be in some sense a kind of physical dimension. That is, just as an event requires three numbers to specify its location in three spatial dimensions and a fourth number to specify its place in time, maybe it needed a fifth dimension to specify its "place in mind." This naive idea turned out be closer to the truth than even I suspected.

As my thoughts took form, I began to suspect that there was a profound and even mystical connection between the psyche and quantum physics. I don't wish to present this connection as a "parallel" or as a metaphor — something I am sure many have thought of or presented before. I wasn't just interested in presenting metaphors, unless the act of creating metaphors provided clues to the deeper connection I suspected existed between the quantum and the psyche.

Observations about the Observer

For the first time in human thought, observation, previously considered the province of psychology, plays an active role in the physical sciences. In earlier times observation was thought to be a purely passive business. One saw, perhaps poorly, what was out there, preexistent. But quantum physics has changed this. Projection is no longer just a Freudian idealistic concept; it has a physical basis. As a result of projection, for example, in the observation of light as a wave or a photon, a small part of the physical world undergoes radical transformation. When anything happens to matter, we must be in-

6

volved. Physics dominates our consciousness even if we don't realize the truth of this.

Can conscious experience be understood in terms of quantum physics? Is this the correct path to follow in order to achieve the ideal of total understanding of all consciousness, including emotions and thought?

The illusion that mind and matter are mechanically separate may cause needless human suffering. Grasping for security will never cease as long as we think of the world as only material and fail to realize our own operational thoughts and actions within it. Perhaps the realization that mind and matter cannot be separated — that they are aspects, much like the sides of a single coin, of one and only one greater and yet subtler reality — will enable the twenty-first century to be born in an atmosphere of peace not yet attainable. To this end I look to the new physics to point the way. Quantum physics has the potential to herald the beginnings of humanistic physics. It, more than any religion, can point to a new and deeper sense of the mind-matter connection. It may even explain the ultimate mystery of life — that it exists.

All physical, religious, and psychological differences may eventually be seen as aspects of the laws of quantum physics. This may lead to startling new realizations about feelings and energy, intuition and wavelength, sensation and physical location, thought and time. If you think, you experience time. If you feel, you experience energy. If you intuit, you experience wavelength; and if you sense, you experience space. That is, sensation and thought are internal dimensions just as space and time are external dimensions. Feeling and intuition are internal qualities just as energy and wavelength are external qualities.

The idea that there may not only be a physics of consciousness but a physics of emotional states, psychological states, altered states of awareness, God, and the outside world is perhaps too much for any of us mortals to bear!

A Round Trip

The study of the physics of consciousness has occupied me since 1971. It began in that year when I took a sabbatical from my post as professor of physics at San Diego State University (SDSU), and traveled around the world. I had earned a somewhat respectable position as a theoretical physicist by publishing papers in prestigious physics journals. I had also risen through the ranks to the tenured status of full professor of physics at SDSU in seven years.

Although my career at the university was quite promising, my marriage had ended the previous year and I felt empty and depressed. My life was passing through a series of crises and traumas.

Meanwhile, the world moved along with its materialistic visions and hopes for plenitude; while in Vietnam, a place most of us had hardly heard of, a battle was fought with the participation of a disheartened America.

My own disillusionment with life was not consciously as severe as the American tragedy. With the newly acquired freedom of bachelorhood I wished only to join my fellow hedonistic Americans in play, sexual fulfillment (*Playboy* and James Bond were my models), and all that the good life of materialism promised. When a fellow faculty member pointed out to me that naive American soldiers were murdering Vietnamese children, I was shocked out of my narcissistic lethargy. Somewhere, buried deep inside of me, was a sense of something, perhaps compassion.

I remember feeling stunned by this news. It didn't seem real. How could we, Americans who had earlier fought for freedom, who provided the torch of hope, who had created from our very spirit the new age of free enterprise, and who had subdued the dreaded Nazi menace, how could we do anything wrong? Perhaps the sabbatical would provide me with the opportunity to see the world and find out. In mid-August 1970 I began my journey.

I decided to travel alone. This gave me the opportunity to experience the world without the influence of any friends or lovers. In Hong Kong, during the first phase of this journey, I met an Indian youth who, on learning of my planned visit to Calcutta, insisted I come and stay with him as part of his family. I accepted. This openness of the people I met in India had a deep effect on me. The spiritual enrichment and material poverty in India were in sharp contrast to the reversed situation in the United States.

It was in India that I began to gain a deeper sense of myself and a feeling of confidence that I was beginning a new phase of my life's work. My earliest enlightening experiences took place in India, but not as you might anticipate. I met no gurus; I visited no ashrams; I didn't even know what those words meant at that time. I found instead a warm and sincere human feeling. It is simple to state it. I felt quite simply and strangely at home there, and physics was halfway around the globe from me.

From India I traveled to Israel, Turkey, Greece, and then made my way west, returning to France and England before embarking for the United States and home again to California. I had been gone an entire year, and it appeared to me as if I had entered another world. Nothing was the same as when I had left it.

I had changed. It was as if another half of me had come alive, with physics now the farthest thing from my mind. I had ceased being interested in things and found people and their thoughts more fascinating.

California Dreamers

When I returned to California, I was offered the opportunity to teach a popular course in physics never before taught at San Diego State, "physics for poets." As such, it promised to give to humanities students an understanding of physics without mathematics. I had always observed, through my experience with students outside of physics and mathematics, that the abstract reasoning demanded by these subjects was extremely difficult for most of them to understand. Indeed, many of them had an almost psychotic fear of mathematics and felt that math was simply beyond them. They expressed, of course, great respect for those who had this ability. They also felt that this ability had little to do with the "real" world of blood, feelings, and people.

This feeling among those who have not developed mathematical skills continues today. Indeed, it is "catchy." (I also began to feel this when I was in India. It was as if that other side of me, finding itself awakened, wanted revenge, while the mathematical side had fallen asleep.) Gradually I had begun to feel that physicists were playing dangerous objective games with reality and ignoring the "real" concerns of people to stop fighting with each other.

Faced with this hostile apathy to science in America,[†] which I had perhaps previously and naively never acknowledged or felt, I sensed an opportunity to learn something about the psychology of learning physics. So I decided to teach "physics for poets." The challenge was what to teach in one semester and what, if anything, would be relevant to the physics-fearing students taking the class.

I found out that what was relevant to them was the very stuff they were most afraid of: the concepts of space, time, matter, the invisible and abstract worlds of Einstein, space-time, electrons, atoms, quantum physics, relativity, and electromagnetism. The worlds of the quantum and relativity, worlds that we cannot experience directly with our senses, offered the greatest excitement.

I began to sense that at the root of the fear of physics and mathematics was a complex line of reasoning that tended to block out those things that were felt to be the *most* relevant to life. It had begun to dawn on me that the new physics had immense relevance to the human psyche, and even to the evolution of consciousness itself. Physics, which is really a quite modern way of thinking, is not even six hundred years old. Perhaps the human psyche devised physics and its root language, mathematics, to assist us with the next evolutionary jump.

Mathematical Belief

Today I believe that mathematics and its application to the physical world govern the operations of our psyches. My faith in this belief has led me to write this book. I believe that the laws of modern physics, the laws of quantum mechanics, apply to our psyches as profoundly as they do to the physical world we all inhabit. I believe this more than I believe in any religious or spiritual leader's dream. Science is the answer, provided we learn how to deal with our fear of the seemingly inaccessible world of abstraction, the world of mathematics and quantum physics.

Many believe that God and science are somehow separate, that human feelings and quantum energy states are vastly different things, that the human potential involves a return to leaving our minds at our backsides as we meditate

---

†Paradoxically, in India science was held in high esteem by students.

on the beauty of nothingness, that mystical states are complementary to physical states.

Although this world with all of its problems appears at times beyond anyone's control, there is no way that the laws of physics cannot apply to all of the universe and be excluded from the inner worlds of our psyches. Some throw their hands up in mystical surrender to the "all mighty," claiming that puny humans are helpless.

I think this is premature. It is not realizing that the laws of physics apply to our psyches that keeps us in fear of physics and in the ignorance of darkness and superstition. Physics can be the way out for those brave enough to even think of our minds as quantum "machines."

Search for Mind

The academic mind is stuck in its insistence that physics be applied to the physical, while psychology to the psychical, and never the twain shall meet. Just attempting to talk about the mind meant the brain to most scientists in both these academic realms. But no, the mind and the brain are not the same; the brain is physical, the mind is psychical.

Those that even dreamed of seeing the mind's processes as physical processes would reduce all mental and psychological or subjective experience to so many particles having so much electrical charge, energy, and momentum. They would subjugate the marvelous and rich world of emotions to data points, making every experience a dead mechanical "plop" of a dot of ink on a recording device's well-delineated output graph paper. The world of mind experience, the world of my fantasies and dreams was far richer than that. How could any physics, let alone quantum physics, apply to this rich realm?

For many behaviorists, other psychologists, and neurophysiologists, not to mention physicists, mind plays no role in the universe — at least not in that universe we have all been taught is *the* physical universe. For how are we to account for mind? Where do we expect to find it lurking? It *must* be a convenient by-product of the physiology of the brain. If we only examine carefully the mechanisms of the brain, down to the remarkable electrical and mechanical movements of the nerve firings and blood flows, we shall find the solution of that overwhelming mystery of all mysteries — the mind.

Wonderful, mechanical, pervading all space and all time, with bits of bits of matter dancing endless dances to tunes played by Isaac Newton, James Clerk Maxwell, and a host of behavioral psychologists, physicians, artificial-intelligence enthusiasts, and others, the physical universe responds to, but has no room for, the piper-playing mind. Mind *is* matter, somehow. Physicists will eventually discover mind in bits and pieces, in shards and in cuttings of tissue, blood, nerves, neurons, DNA, RNA, and all out of that lofty sphere that sits atop the neck and shoulders and encloses the human brain.

But perhaps our optimism is misplaced. The mind may not exist in the physical universe at all. It may be beyond the boundaries of space, time, and matter. It may use the physical body in the same sense that an automobile driver uses a car. The mind may not be a function of mechanics or electrical firings of neurons. Yet it could appear that way. It certainly seems logical to think of it as contained in the brain and nervous system and existing totally as a mechanically operating system. To see how we could come to this possibly mistaken conclusion, namely, that the mind is electromechanical in origin, think of a car with darkened windshield, rear, and side windows. No one can see in from the outside. The car moves mysteriously down the street. Is there a driver? Suppose you just descended from the planet Googoo. You have just fallen in the the middle of a Los Angeles freeway alive with purposefully moving darkened glassed autos. Being made of oog-oog, you have never seen a human being before. Your mission (if you decide to take it) is to discover how these unusual "organisms" function. Since Googoo is a very scientific planet, you know what you must do: Analyze. Take the "organism" apart.

First pull off a wheel. Ah! See! It still functions. It looks a little funny moving that way, scraping along on three wheels, but nevertheless it still moves, still seems to have a mind of its own. It's just a bit "physically handicapped." Oh, look! Under that front lip, which opens up, there is something that makes a loud noise. It says "motor" on it. And look at all of that electrical activity. Lets pull a few wires. There! That did it. It stopped. We have finally discovered what makes it tick. We have found the "creature's" mind.

Clearly, if we remove any of the parts of a brain, the brain and body will not function very well. But we would feel quite justified in thinking that we had discovered the mind, or at least the physical-mechanical cause of it. But modern physics, particularly quantum mechanics, has made us realize that the universe is not mechanical after all. It doesn't consist of solid bodies filling space and taking up time. It has properties that are, as amazing as it may at first appear to our Western minds, more like *ideas* than physical objects. Just as a simple change of your mind wipes the slate clean for the next thought, a change of a physical apparatus alters the physical attributes of a solid object so that the next experiment produces results that have no logical connection with the previous one. Physical properties of atoms and molecules depend on the choices made by the observer of those physical properties.

Where does this leave our minds? Nowhere — at least nowhere in the physical plane of what we call existence. Instead of being a product, mind may be the creator of the physical world. Or it could be a process parallel to physical movement. Weaving and woofing, like the threads of a complex Turkish carpet or Buddhist *yantra,* through countless brains, human and otherwise, mind designs itself into bits and pieces, separate existences, so that it can fool itself into believing that the physical world really exists separate from itself. What a wonderful illusion! But for so expensive a price. Insanity, depravity, starvation, war, and apparently endless suffering accompany beau-

11

ty, love, sanity, bliss, and excitement.

If we think about it at all, even for the briefest moment, we must marvel at even the simplest recognition we can achieve. We are able to form thoughts, create pictures that we see somewhere (presumably that somewhere is in our heads), construct through our vocal mechanism a complex coordinated orchestration of sound and thought (which although musical is rarely appreciated as such) that we call words, and — best of all — we can use this musical thought as a rather sophisticated form of communication. We have learned to do this so automatically that few of us ever question how we do it.

Many thinkers, writers, and philosophers believe that such abilities will never be completely understood. Others feel that the process of understanding this ability will come from a deeper study of the brain and its mechanisms. Still others think that this ability is totally God-given and therefore incapable of being understood. The list of theories and ideas about how we think is perhaps endless, stretching back into time to the earliest human observer.

I believe that modern physics, particularly quantum mechanics or quantum physics, can not only shed light on the mystery of this most obvious human ability, the ability to think, but can solve the mystery of the mind itself. That is, quantum physics provides a theoretical basis for understanding the mind's basic functions: intuition, feeling, sensation, and thought.

PART ONE

$$\Psi * \Psi$$

QUANTUM PHYSICS
AND THE MIND OF THE OBSERVER

To probe the psyche we need all the tools we can find that have been invented by the most creative, analytical minds. We have had the Einstein, the Bohr, the Planck, and the Heisenberg minds to start us along the path to a new synthesis. The next step is the Heisenberg of the mind itself.

1

Space, Time, Matter, and Spirit

If you think, you experience time.
If you feel, you experience energy.
If you intuit, you experience wavelength.
If you sense, you experience space.

Mind and the "New Physics"

When the idea first occurred to me to write this book, I began to wonder to myself why I ever thought mind and "new physics" had anything to do with each other. After all, physics deals with the external, objective world — the world "out there" that confounds, frightens, and sustains us — while the mind is internal, subjective, and "in here." Yet, the mind can confound, frighten, and sustain us, too.

But the mind is elusive. We can't pin it down or take off its wings to see if it can still fly. We can't put it into a linear accelerator and observe what happens to it when bombarded with 100 MEV protons. Indeed, there appears to be no way to experimentally study the mind. This has led many to believe that the mind simply doesn't exist. Behaviorists, for example, believe that mind is a by-product of material phenomena and that thoughts are just so many rearrangements of molecular groups. These rearrangements, they theorize, affect the amounts of neurotransmitters emitted or created at the presynaptic end of a neuron. Thus thought is reducible to chemical reactions and memory becomes a storage problem in the synapses, where the amount of transmitters is affected by serotonin and cyclic adenosine monophosphate (AMP).

It is natural that such theories should be formulated. Cyclic AMP and serotonin released in neurons can be studied. We can inquire into what causes these chemicals to be produced in neural cells. However, ultimately this reductionist program leads to a boundary that cannot be experimentally observed, to the "twilight zone" of quantum mechanics. At some future time we will know what the mechanisms are for producing memory. What we won't determine without quantum physics is why some cells remember and some don't, and why some members of some species survive and change and why sisters and brothers of the same species do not or "will not."

Somewhere, even at the level of the simplest biological cell, such as is contained within the marine mollusk *aplysia californica,* the element of chance enters. This element, however, does not just appear randomly but as opportunity. A minuscule form of "will" is manifested. A sense of the past, present, and future appears. A behavior modification takes place in which a symbolic reference — a kind of shorthand code for experience — takes shape. A sense of identity forms. In this primitive beginning we are, I propose, looking at the prime effects of the "new physics" — the appearance of consciousness.

These effects cannot be based on already existing material causes. They are instead based on nonmaterial elements, those that a quantum physicist must deal with when he or she calculates the probabilities of events occurring. Quantum probabilities, the kinds of probabilities determined by the "new physics," are not simply statistical enumerations and arrangements of elements too numerous to count otherwise. It is not just a case of: "How many ways are there to put four coins in a box with only two heads showing? Now repeat for n coins with m heads showing."

The reason is that in quantum physics the elements are not physical themselves; they do not exist as objects. Their very existence depends on the idea of their existence beforehand. They are treated as "tendencies to exist" rather than as already existing possibilities like the sides of a flipped coin. In the quantum world the quantum coin's sides do not appear unless someone calls for them to appear.

A typical example is the molecular structure of the ammonia molecule NH_3. It resembles a pyramid with the three hydrogen (H) atoms forming an equilateral triangle and the single nitrogen (N) atom forming its apex.

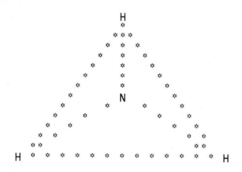

Figure 1
Top or bottom view? A quantum picture.

The trouble with this picture is that the molecule vibrates. It won't stand still. This is due to the uncertainty principle of the "new physics," which states that all matter must undergo this unruliness because it is not possible for the location and the momentum of any object to be specified with arbitrary accuracy. This means that the NH_3 structure must allow itself to move so that its atomic elements are not "pinned down." By a careful calculation we find that the N atom can "appear" to be on either side of the triangular base of H atoms. Thus in the above picture you can imagine that you are either looking down on the apex or looking up at it through its base of H atoms.

Surprisingly we find that in order for the molecule to be existing in a stable energy state — one in which its energy is not changing — the N atom must be undergoing a dance in which it is, "so to speak," in both positions at the same time. The uncertainty principle allows the molecule to have a stable energy if it abandons its stable or fixed configuration in space. And vice versa, if we fix the N atom above its H atomic base, the energy of the molecule is no longer constant.

Thus we calculate the energy of the molecule only if we allow the position of its apex to have equal "tendencies" to exist above and below its triangular H atomic base; and we calculate the position of the N atom above its base only if we allow the molecule to have approximately equal tendencies of being in either of its first two energy states (the ground and first excited state). Which is the ultimate truth? The answer depends on what we are looking for.

This freedom of choice of ours applies all the way down to our atomic constituents. We choose and the world appears, like Aladdin's genie, who comes when summoned from the lamp. The only problem is that we do not know what we will get as a value or specific content for the attribute we choose to have appear. In the example above, I wrote that if we choose to calculate the position of the N atom, we lose any evaluation of its energy. This is true. But in choosing position as a desired quality we must put up with the possibility that the N atom could "appear" below the triangle just as well as above it the first time we choose to observe its location in space.

By repeating our measurements carefully we will eventually prepare the molecule so that it will always be "apex up." The minute we attempt to measure its energy, this preparation will be destroyed. It will appear with one of its two most probable energies. With further repetition one of those energies will appear repeatedly. However, no "memory" of its previous apex position will exist.

Thus the simple molecule's behavior depends on how its environment "looks" at it. It is not just a subunit to be plugged into a device. The consequences of the disappearance of energy or positional configuration are not just those of a passive fragment that can be fitted into the larger puzzle in two different ways. The potential aspects of quantum physics lead to infinite possibilities depending on all kinds of compromises that can exist in the environmental "question" put to the molecule. "Are you energy or are you in position?" can have many answers depending on how much we narrow the accuracies of our measurements. By releasing a little control in how we determine the energy, we gain a little control in what configuration the molecule is to be found.

Thus we conclude that the "new physics" introduces the element of consciousness into the material world. This consciousness will not arise from the molecule itself, as seen as a material unit, but will arise as a "risk-taking" psyche — that is, one that chooses. These choices cannot be made willy-nilly. "Reason" must begin to make its appearance, which surpasses the simple mechanism of cause and effect. We know that atoms do not follow the laws of cause and effect except statistically or on the average. To explain the evolution of learning, associative memory, and possibly even the more primitive forms of memory called habituation and sensitization, we must face the quantum. States of consciousness, feelings, emotional states, and psychology as a science may depend on the recognition that mind, the consciousness of the universe, arises through quantum physics.

The "New Physics" of Time

Perhaps the greatest mystery of all time is "time." There has never been an adequate definition, a clear metaphor, or even a good physical picture of what time is. As far as a physicist is concerned, time appears as a parameter, or independent variable, in the equations of mathematical physics. Jean Piaget points out that young children grasp the notion of time sequence in an operational, nonintuitive manner. Until about seven or eight years of age, children do not possess the ability to reason about several possible alternative temporal sequences. Instead they tend to confuse concepts of time with those of speed and distance.

It's not only children who are confused. In quantum mechanics time is not even an observable! Time appears in the equations of quantum physics as an extraneous ordering parameter. It seems that there is no way to talk about time without talking in circles — introducing other concepts such as causality, order and disorder, the second law of thermodynamics, entropy, vibration, motion, history. And these concepts in turn depend on a concept of time.

In working with the experience of time, I am aware that part of the confusion is that time in and of itself is a relative concept. To illustrate, I cite a familiar Zen Buddhist parable:

One windy day two monks were arguing about a flapping banner. The first said, "I say the banner is moving, not the wind." The second said, "I say the wind is moving, not the banner." A third monk passed by and said, "The wind is not moving. The banner is not moving. Your minds are moving."

In the first monk's view the movement of the banner is observed relative to the monk. The monk's view of himself is nonexistent, a typical classical-physics view of time and motion. The role of the observer, even his presence, is nonexistent. The wind, being invisible and thus having no substance, also cannot be moving. The second monk sees the banner as under the direction of the wind. It has no cause to move. The wind is the true mover. The banner thus remains where it is. It ripples, but only in the wind. The third monk, however, poses the problem of relativity and reawakens us to ourselves. Nothing out there moves, he states. It is only our mind passing over the landscape of the "out there" that moves.

This relativity is commonly experienced in motion. Sitting in your car at a stoplight you may have had the eerie experience that you were rolling backwards with the brake well set. After ascertaining that all was well and that you were at rest, you probably noticed that the car alongside your own was creeping forward, edging ahead to get off to a fast start. So from one viewpoint you moved and the outside world stood still. From another viewpoint you stood still and the outside world moved by. This experience of movement is commonly associated with the experience of time, but this is an error. The time experience is *not* an observable!

The closest we come to observing time is observing what the Buddhists call "being-time" as described by Dōgen Zenji. The closest I can come to describing this experience is to say the word *here*. The experience of time is the experience of "hereness." Time has nothing to do with movement. Things move, but time stands still. When we say time passes, we mean that we pass. When we say we observe the passing of time, we are doing no such thing. We are observing the passing of our minds, the "movement" of our thought processes.

The notion that things move and my mind (whatever that is) observes the motion carries with it the sense of *impending,* which is the root cause of fear. This sense is created through the feeling that past, present, and future are not the same. What is past is literally gone, dead, vanished, and never to be seen again. What is the future is also not here, not experienced except as a hope or a dread. Both past and future are not here! Our experience seemingly is reduced to now, but that too is an illusion, for how now is now? To say "now" and to know I am experiencing "now" requires something beyond now; it requires both past and future. The word *now* cannot even be heard now because to hear "now" requires the length of time it takes for the consonant and vowel sounds to register in our brains. Thus "now" is also a phantom! "Now" is not here. It cannot be experienced as time.

Thus time as an experience, that is, *time itself,* has nothing to with past, present, and future. These are events in time but not time itself. Time as an experience in itself is paradoxically — in the sense of ordinary time in terms of past, present and future — timeless. That is, time takes no time. Time is thus primal and not secondary. It cannot be derived from anything else. It is not, again, motion.

Being-time is different. It is the sense of here. It is our sense of presence. It does not need space, feeling, knowledge, sensation, emotion, or others to be experienced. Here needs no object for itself. Here is the being-time of Zen Buddhism. With this intuitive concept of time in mind we can talk about all that is not being-time. Suddenly the third monk's viewpoint takes on a new light. Everything that is, is, was, and will be. It remains "out there" forever. Things do not pass away in time. Every moment remains lifeless, motionless and frozen forever. We — that is, our being-time — sweep across the landscape of all experiences as an airplane passes over the Grand Canyon. Past, present, and future represent a map for the perusal of the all-seeing being-time. The classical physical concept of time, what we call time, then, is nothing more than a coordinate on this map, a parameter.

According to the "old physics," time is a parameter. This means that the qualities we observe, such as motion, energy, and location of objects, are "mapped" according to a value assigned to the time parameter. Thus at $t = 3$, the ball is on my knee. At $t = 4$, the ball is on the floor. We say that "time" flows from 3 to 4 as the ball falls from knee to floor. We draw the map as in Figure 2.

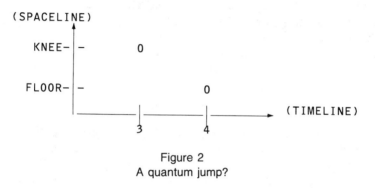

Figure 2
A quantum jump?

How the ball got there is another matter. We say that the position of the ball depends on time. This dependency arises from the law of, for example, falling objects, known as gravity. In this case we would draw a parabolic line through the two points (O) on the graph. We would then think to ourselves that the ball fell continuously in time.

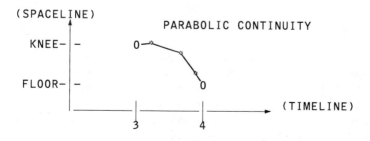

Figure 3
Or a continuous movement?

It first came as a shock to me that time does not have a mathematical, quantum-mechanical operator representing it. Not to have an operator representation in quantum mechanics is serious business. If a certain concept is determinable as a number — in other words, has a measure according to a scale by which it can be compared objectively — it must be representable as an operator in quantum mechanics. If no such operator exists, no such observation is possible. That is, in spite of our commonsense notions about time, it cannot be observed. Of course we all carry watches and we all watch the sunrise and sunset and set our clocks back or forward one hour on the appropriate days of the year. Yet in spite of all these common measurements of time, it is not observable.

What are we observing in all of the above? The answer is motion. We infer time as a result of motion. We compare the movement of a vibrating spring to the rise and fall of the sun. We infer time by comparing movements. The sweep second hand does not sweep time; it sweeps space periodically. That means it sweeps space in time.

The "New Physics" of Space

Captain James T. Kirk aboard the starship *Enterprise* opens the show, saying, "Space, the final frontier . . ." Since the early sixties we have become "space crazy." Space is all that is "out there," waiting to be explored. Space is what we live in, sleep in, and drive our cars through. We fly through it. We move in it. We occupy it as if it were ours to occupy.

In dealing with time I emphasized that we needed to consider what we meant by the experience of time. Even that was difficult to talk about since it could be interpreted in different ways. Finally I had us consider the concept of being-time. This, I claimed, was not the experience of motion, nor the experience of past, future, or even the present. I claimed that paradoxically being-time was the experience of "here." I also explained that "here" cannot be represented by a quantum-mechanical operator. In other words, "here" is not an observable; we cannot observe "here" because the act of observation requires an observer and an observed. It needs separation to exist. That is why "here" cannot be experienced without there being a "there" — in other words, a "now."

"Now" is an observable. "Now" is the experience of space. When we say "now" we are not experiencing "here"; we are experiencing what we have come to call "space." Looking out at all that is around you, you experience space. You experience this feeling or sensation of space now. You certainly don't experience it before or after now. You have it now. "Now" is therefore the simultaneous experience of many "heres." Only these "heres" are now "theres." They are "out there" in the world.

When we say we are measuring space, we are simultaneously measuring two or more locations. A measure of the size of an object would not make sense if we measured one end of it at noon and the other end at midnight. The object could have moved in the meantime. Thus the measure of space implies the single instant. It implies what we call "now." "Now" equals space. "Here" equals time.

Whereas time could not be an observable (later we will say what it is), space is an observable. We know how to measure it and how to determine its properties in the form of dimensions. We have learned to analyze motion as degrees of freedom in space. We have discovered that space is "cubical," consisting of three spatial dimensions, length, width, and depth, or up-down, right-left, and front-back. We need two "directions" for each dimension. Up is opposite to down, right is opposite to left, and front is opposite to back. We see that we can move in either direction in any dimension. Thus space is an observable because we can freely operate in it. That which we can operate in is operational. That which we can observe operationally is observable.

To observe space we need the observer and the observed. Their separation is "space." Their separation is the "now" experience of the observer. Using a geometrical analogy, "here" is a point and "now" is a line, a surface, or a whole solid. All lines, surfaces, and solids can be generated from a point. Thus space is generatable from time. The observed then equals space. The observer equals time.

Quantum mechanics has no problem with space. Space is an observable, which means that there exists an operator to measure space. When we use this operator in the world, a number specifying the location of an object is produced. We say the object is "there."

This experience of "thereness" I call *now*. It is, however, ultimately our own individual experience — the experience of "here" (time) — that is responsible for our "now" experiences. All objectivity ends in subjectivity. Thus we see that all experiences ultimately are reduced to "here." All experiences are only experiences if they are "here" experiences. This includes the spatial experience of "now." That is why when we say "now," we usually mean the experience of time even though we are "looking out" at the world when we say "now." It is here that the *here and now* are entwined. There cannot be a here without a now. The two are inseparable. This is the experiential basis of relativity.

We might say that this is the temporalization of space in contrast to the physicist's usual way of spatializing time by drawing a timeline on a sheet of paper and saying that time is the fourth dimension. In recognizing this we come to the conclusion that time is the observer. Space is the observed. That is why time cannot be observed. The observer is invisible. That which is observed becomes space. By temporalizing space we recognize consciousness or mind as "out there." By spatializing time we deny that recognition. We make time "motionless." We freeze it. We then lay it out as explanation, the planing of experience on a sheet or roll of paper. We write it out. We symbolize it. We make it cause and effect. We turn life into mechanics.

A SPACE-TIME DIAGRAM: THE MAP OF EXPERIENCE

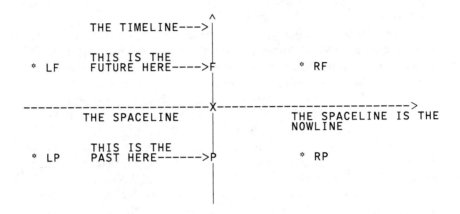

Figure 4

The above shows a space-time diagram marking five events separated in space and time as witnessed by one observer. The vertical line is called the timeline and represents the history of this single observer. The horizontal line is called the spaceline and represents different locations to the right and left of the observer that are the positions of events happening simultaneously with the single event marked X (located in the center).

X marks the spot of consciousness where the observer experiences at the time "here" the space around him "now." He actually cannot experience this space at all because it takes time for signals from all the points on the now line to reach him. He "thinks" now when he looks out at space around him. P and F mark time points in the past and future, respectively. The event *LP happened in the past and to the observer's left. Event *RP in the past to the right. Event *LF will happen in the future to the left. Event *RF in the future to the right.

Space-time and the Simplicity of Einstein's Relativity

Thus we are led to space-time. Space without time is unobservable because there is no observer. Time without space is also unobservable because there is nothing to observe. Time becomes what we call pure consciousness. The game of this consciousness is recognition and correlation — the building up of a consistent picture of the universe. It does this by putting things in order using itself as a reference point. This mapping process I call the spatialization of time. It is pure observational procedure. It is what observation does.

Its success is, however, its own downfall. By "understanding" the universe, consciousness retreats and disappears. It vanishes into so many boxes labeled

"DNA," "RNA," "cyclic AMP," and "the material of the brain." But there is another process going on. It is a process that many would falsely label as mystical or spiritual in the derogatory sense. It is the recognition that consciousness belongs in the universe. Consciousness is and is not derivable from matter. We could call this process the temporalization of space. It is what many mystics call "letting the observer become the observed."

Einstein made the maximum possible use of the idea of spatialization of time. By recognizing that the transformation of coordinate bases known as the Lorentz transformation had to include the time coordinate of one base in the specification of the space coordinate in the other, Einstein made us realize that time and space were intermingled. But he didn't go far enough.

With the help of Hermann Minkowski, relativity theory became a theory of geometry. It was all spatial. Time became a spatial coordinate. By seeing time as space, physicists developed a picture of the universe that was observer-independent. The observer was gone. He or she was put on a sheet of paper and turned into a timeline. But the experience of observing can never be mapped. Observation is not an observation. Experience is not an experience. Time is not space.

THE SPATIALIZATION OF TIME

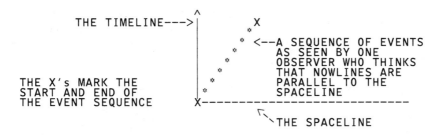

Figure 5

Einstein's space-time is observer-independent. It shows us how events are mappable in a democratic, observer-independent manner. It brings out the invariance of relationship that exists between events. It helps us to understand how two different observers could determine two different measures of the time and space coordinates of two different events. It says nothing about how one observer observes the time and space coordinates of one single event, any single event at all. It says simply that there is a mathematical relationship, which is no more complicated than the theorem of Pythagoras, between the events as observed by two different observers. If we map the events as points on a plane or points in a space of three dimensions or, ideally, as points in a four-dimensional space, then the events will lie on a straight line in that space. With any straight line it is possible to form a right triangle in space with that straight

25

line as its hypotenuse. It is possible to form a whole family of right triangles with the same hypotenuse. In the space of space-time the Pythagorean theorem is still true with a slight difference.

The only difference is that since we are viewing things in space-time and not in space, right triangles will only look like right triangles for one observer, the one who draws the map with the timeline vertical. For him time moves up the page while he sits still at one location in space. From his point of view a second observer is one who is moving relative to him. She appears, for example, to be moving toward the left. According to relativity, her perceptions of here and now are not the same as his. Although she would see her own timeline crossing her spaceline at a right angle, he would not see things this way. This mapping distortion does not play any role in the mathematics. It is nothing more than the kind of distortion one must have in attempting to flatten out what is made in more than two dimensions. This is, of course, common in ordinary mapmaking. The flat Mercator projection map distorts lines of longitude by making them parallel where in truth the lines converge at the poles.

The Pythagorean formula relating the sides of a right triangle, *A* and *B,* to the hypotenuse, *C,* is simply

$$A^2 + B^2 = C^2$$

Just as it is possible to walk north and then west to reach a destination, it is also possible to walk southwest and then northwest and reach the same destination. As long as the distances walked are the sides of a right triangle, the distances walked off will follow the Pythagorean formula even though the

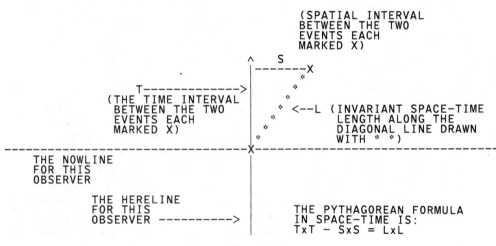

Figure 6

separate legs walked off will not be necessarily equal to each other. In a similar manner time and spatial intervals form the legs of a right triangle in space-time. This means that two events in space-time are always separated by an invariant diagonal "hypotenuse" called its proper length or proper time. Observers "walk off" different legs of time and space according to how they are moving. Thus one observer might see a time interval T between the two events while a second observer would see the time interval T'. Einstein and Minkowski showed us how to reconcile the differences through use of the space-time Pythagorean formula

$$T^2 - S^2 = L^2$$

T and S are the right legs of the space-time triangle and L is the hypotenuse.

THE PYTHAGOREAN THEOREM IN SPACE-TIME FOR A SECOND OBSERVER

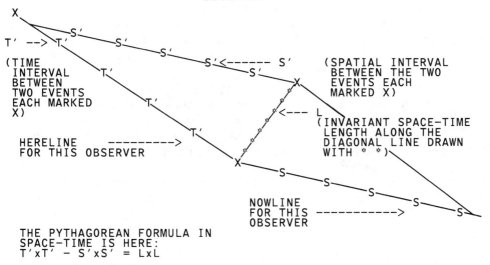

Figure 7

Notice that the two formulae are identical except that in space-time we use a minus sign while in space we use a plus sign between the squared legs of the triangle. The minus sign appears counterintuitive to us because we are dealing with the geometry of space-time and not space alone. Time is not a real space dimension. One could even argue that space-time is not a real space and the space-time Pythagorean formula is not a real right-triangle formula. Such arguments imply that space-time is an analogy of space.

However, we could, from the point of view of space-time, just as well argue that space is an analogy of space-time. This is perhaps the view of Einsteinian

27

thinkers who hope to geometrize the universe. In their vision all events are spatial-temporal in the geometrical or stagnant sense. This could be viewed as the view of physics.

In such a view time is "layed out" for all to bear. Experience is entirely mappable. The territory is the map. Or, as I put it earlier, time is spatialized. Experience becomes "dead," mechanical, and "layed out" as cause and effect.

The goals of psychology are perhaps different (or I should say that the goals of the psyche are different). The psyche, insisting on its existence, its reality, its experience, its "I-ness" hopes that physics is wrong. It thinks, senses, feels, and intuits that it is wrong. Its goal might be stated as the temporalization of space, the bringing into consciousness the inert world of matter that seems to exist "out there" separated from us all.

The "New Physics" of Matter

For several years now I have been involved with various departments of education, particularly in California, as a consultant working with gifted children. My interest has been in communicating the ideas of the "new physics" to children, particularly those in the fourth through sixth grades. Instead of the usual approach of demonstrating a relationship that appears to exist — say, the relationship of color to the use of filters, or the speed of a falling object to the distance it has fallen — I have attempted to "bring children to their minds." I feel that children like to wonder, given the opportunity and the safe environment to do some speculation and imaginative guessing. I often perform magical tricks with coins, cards, and ordinary objects to convey the magic inherent in the "new physics."

Inevitably, in their desire to catch me at my trickery, the children come up to the table to explore and observe just how the trick works. I rarely show them how I do a trick. However, my purpose is not to deceive them. It is to make them think logically and intuitively and to come to their senses. By attempting to figure out what I do in a trick they are using the same type of reasoning that a scientist does.

In one of the demonstrations I show two small bags, one red and the other yellow. They are of equal size but not of equal weight. One actually contains a lead ball and the other a plastic ball weighing about one-eighth the former. The problem is to determine which one weighs more without lifting the bags off the table. Think about it for a while. How would you prove or demonstrate that one bag is heavier than the other without actually lifting them up and holding them in your hands?

No, you can't put them on a scale. No, you can't open them up and peek inside. No, you can't smash them on the table. After a while someone asks if they can be pushed off the table and allowed to fall to the ground. "Yes," I say. But a few children recognize that this will get them nowhere. All objects fall with the same acceleration — that is, at equal times during their fall the

28

objects would be at equal heights from the level ground — if they were both released at the same time. This was the great discovery of Galileo. Indeed, I pick up the objects and drop them together. They hit the ground at exactly the same time as far as the eye can tell.

There is a seeming magic in this. Why should two objects weighing differently fall alike?

Finally someone wants to touch the objects and, usually with a little encouragement, shake them around on the table. It soon becomes obvious: The heavier object resists motion more than the lighter one; it is more difficult to move it about.

Thus we come to realize that weight is not a measure of this resistance. Since the child shook the objects on the table, she was not weighing them. Yet she determined their relative weights. The resistance she experienced is called *mass*. It is equivalent to resistance. It would be felt in each and every single object even if the observer and the object were floating in free fall in outer space. The more mass, the more weight there would be if the object were put in a situation where its weight was determinable, as in the gravitational field of another object.

But what gives mass its properties? Our ordinary senses tell us that mass takes up space and moves in time. Mass is inert, from which the notion of inertia arises. Inertia is just another word for this tendency of the object to persist as it was. Our efforts to move it about are met with its inertia. A new insight into the experience of inertia came about because of Albert Einstein. Einstein saw that inertia was energy.

Einstein, using concepts of classical physics, including energy conservation, came up with a brand-new form of energy. This was the potential energy contained in an object and represented the object's mass. The formula $E = Mc^2$ describes energy E as contained within mass M, which when released or transformed would have magnitude M times the speed of light squared, c^2. Einstein wrote:

If a body gives off energy L in the form of radiation, its mass diminishes by L/c^2. The fact that the energy withdrawn from the body becomes energy of radiation evidently makes no difference, so that we are led to the more general conclusion. . . . The mass of a body is a measure of its energy content.

How does energy become mass? Or mass change into energy? Einstein, in his original work in 1905, felt that it was possible to observe matter changing into energy by measuring radioactive radium salts, which were known spontaneously to emit nuclear radiation. A minute amount of matter, M, provided a great amount of energy, E, because the speed-of-light factor c^2 was so enormous. If all the mass of an atom was "atomized" into light energy, it would cause a great deal of damage. The mere transformation of the mass of a single brick of concrete produces enough energy to raze a city with a million

29

inhabitants. Even Einstein failed to realize, as late as 1921, that "atomic energy" was releasable in such enormous quantities. This destructive element came later.

For Einstein there was the "understanding" that mass and energy were equivalent. If an object was increasing its speed as it moved along, it had to be increasing its energy and therefore its mass. But when that object was laid to rest, it still had its original "rest" mass that it had all along.

Matter was somehow "stuck" or unavailable energy. Mass and energy were equivalent, but each satisfied its own law of separate conservation in the sense that the "rest" mass remained as a tough nut on both sides of the energy conservation law.

But where does mass get its "stuckiness"? Just for a moment, go on an imaginary journey with me as we push on a tennis ball weighing, to start with, 4 ounces. If we imagine that we can push on this ball so that a flea riding in the ball feels a steady acceleration of 1 g — that is, the acceleration of gravity as felt on the earth; and so that the flea experiences one year of life aboard the tennis-ball spaceship, then according to relativistic physics the ball's mass will have increased to more than 6 ounces. It will be traveling through the universe at more than three-fourths the speed of light and it will have traveled more than one-half a light-year (the distance light travels in a year). The journey will appear to take just over a year's time.

If we allow the flea to ride for fourteen of his years, he will get quite a trip. His journey is shown in Table 1.

By the end of five years the flea will have reached .999909 the speed of light, have gone over 70 light-years from earth, and be moving on a ball weighing over 18 pounds. By the end of 10 years' travel he will have reached over 10,000 light-years and his spaceship will appear to weigh nearly a ton and a half. By the end of the journey, although the flea will have aged only 14 years, nearly 60,000 years will have passed on earth and his tennis-ball spaceship will weigh more than 150,000 pounds, or 75 tons!

As the spaceship continues to accelerate, its speed hardly changes. All of the energy expended to speed up the ball is going into the ball's inertia.

However, all of the above has been determined from the vantage point of the earth. Aboard the tennis-ball spaceship all is perfectly normal. The flea is indeed enjoying his ride and feeling the effects of a normal 1 g gravitational field on his body. It is only to us back on earth that the ball takes on such grotesque characteristics. In other words, the inertia of the ball depends on its energy, which in turn depends on who is looking at it. The surprising thing about relativity is that the speed of the ball is not changing even though the ball is undergoing acceleration. This acceleration is felt at each and every instant by the flea. We back at home will not see any increase in speed. As far as we can tell, the ball is moving at a little below the speed of light.

Thus in some way the inertia of any object — that is, its mass or substantiality — depends on consciousness. It depends on observation. Carrying this

A FLEA'S JOURNEY AT ONE "G"

Table 1

S = 0	X = 0	T = 0	V = 0	M = .25
S = 1	X = .526788	T = 1.13995	V = .761594	M = .38577
S = 2	X = 2.67933	T = 3.51805	V = .964027	M = .940548
S = 3	X = 8.79563	T = 9.71734	V = .995055	M = 2.51691
S = 4	X = 25.519	T = 26.4712	V = .999329	M = 6.82743
S = 5	X = 71.0137	T = 71.9771	V = .999909	M = 18.5417
S = 6	X = 194.694	T = 195.662	V = .999988	M = 50.4286
S = 7	X = 530.898	T = 531.867	V = .999998	M = 137.079
S = 8	X = 1444.79	T = 1445.76	V = 1	M = 372.62
S = 9	X = 3929.02	T = 3929.99	V = 1	M = 1012.89
S = 10	X = 10681.9	T = 10682.8	V = 1	M = 2753.31
S = 11	X = 29038	T = 29039	V = 1	M = 7484.27
S = 12	X = 78935.1	T = 78936	V = 1	M = 20344.3
S = 13	X = 214570	T = 214570	V = 1	M = 55301.7
S = 14	X = 583262	T = 583263	V = 1	M = 150326

S is the time in years the flea is moving.
X is the distance in light-years the flea has traveled
    as determined by Earth.
T is the time in years the flea has traveled
    as determined by Earth.
V is the speed of the tennis ball in terms of
    the speed of light.
M is the mass of the tennis ball in pounds.

further, consider a ball on a string being held by a child and swung around his head in a circle. The child holds one end of the string and the ball is whirling around above his head at the other end of the string. As the child swings the ball faster and faster, the string must be held tighter and tighter because the ball is undergoing acceleration and, according to Newtonian mechanics, the tension in that string is the cause of that acceleration. The greater the tension, the greater the centripetal acceleration, i.e., the faster the ball will move.

But our friend the flea lives in the ball. He is oblivious of children and strings. He only knows what he feels and he feels weirder and weirder as the ball is accelerated. From his point of view he feels the inside wall of the ball pushing on him (he is resting on the side of the ball opposite the string). He concludes that there must be a force acting on him from the outside world that is pushing him against the ball's inside (opposite the direction of the tension in the string, which he knows nothing about). He senses that the force is increasing and it worries him a little.

Since he is feeling heavier and heavier (a similar experience is felt by jet flyers when they "pull g's" as they come out of a dive) the flea gets the impression that his inertia is increasing. This force of inertia felt by the flea is called the inertial force, or centrifugal force. He feels it because he is riding in a ball that is accelerating as the boy spins it faster.

31

But what is the ball accelerating from? The boy? The earth? From the boy's point of view, the ball accelerates with respect to him. He is at rest. He is in what is called an "inertial frame of reference." The ball is not. It is in a "noninertial" frame. That is why the flea feels the force he does.

If the boy now attaches the loose end of the string to a ceiling beam and sets it in motion, it will swing like a pendulum. If he leaves the ball swinging and returns several hours later (assuming that the ball is able to swing that long), he will find that the plane in which the pendulum is swinging, its so-called inertial plane, will have rotated with respect to the ceiling beam. This is the well-known Foucault pendulum experiment, which demonstrates that the earth rotates.

The ball maintains its inertia. It keeps swinging in the same plane. The earth rotates. This changes the angle of the ceiling beam (attached firmly to the earth, we hope) with the inertial plane.

Inertia persists! It isn't even touched by the rotation of the whole earth! But with what frame of reference does the plane align itself? I mean, how does the ball "know" not to rotate with the earth? Newton thought that the ball received its instructions from the infinite, from God's reference frame.

If the boy now looks out at all those stars in the universe, he will see an equally amazing demonstration of rotation. The stars at night are big and bright and rotate around him at the plane rate. In other words, the earth is rotating with respect to the big bright stars, galaxies and galaxies of galaxies with the same rotational speed as the inertial plane of the swinging ball rotates with respect to the earth. The inertial plane of the ball is "following some unknown instructions" from the stars at night. Inertia depends on all those distant stars more than it depends on the earth, the sun, the planets, and all our nearest neighbors in the sky. Shades of cosmic astrology!

Mass is resistance to change. This resistance is an outgrowth of the whole universe. If the universe were to somehow disappear, leaving only you in it, your inertia would also vanish. You would be a ghost. Without the universe there is no you, at least no you capable of experiencing matter.

The Speed of Light as an Upper Physical Speed Limit

Going back to the tennis-ball spaceship and realizing that to the flea everything is normal, even though he is moving at almost the speed of light, we realize that the inertial content of the ball is dependent on the reference frame in which it is seen. We also realize that the ball can never reach the speed of light because increasing its energy goes into increasing its mass, not its speed. This is true for anything that has inertia. Thus matter is limited to light-speed minus a tad no matter where or how far it has traveled.

The upper bound of light-speed and the inertial content of matter are connected. This limiting speed plays an even more significant role. Nothing

THRee
THEORieS:

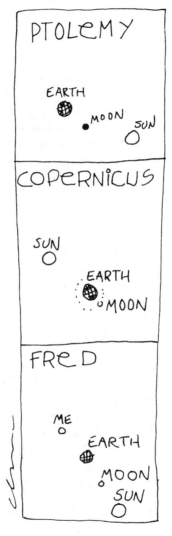

physical, not even a wave of light nor sound carrying information, can move faster than light.

When I was a child, my friend, who happened to live in the building next to mine, and I hooked up a string telephone between our apartments. By taking two empty tin cans, puncturing holes in their bottoms, pushing the ends of the long string through the holes and knotting them, and finally giving my friend one of the cans while I held the other, we were able to talk to each other while

I was home in my apartment and he was home in his. Sound communicates. It moves at the speed of sound along the string. We could whisper to each other and our parents would hear nothing. To us, it felt as if we were communicating instantaneously. Our apartments were only 10 or 12 feet apart, however.

Another day my father, who loved to play golf, took me on a golf outing. I remember my feeling of awe while standing on the green preparing to putt as I watched the members of the party behind ours tee off towards our green. Since we had already reached the green and since the course was crowded on that spring day in Chicago, we politely waved the upcoming party on. I saw each person hit the ball. I watched carefully as the ball flew through the air (so that I wouldn't get hit in the head), but I heard no sound. It came a second or so later, after I saw the player hit the ball. It took me some time to realize why I hadn't heard the smack of the golf club on the ball: Sound information traveled slower than light did. It is a ghostly feeling to watch a golfer swing at a ball, hit it so that it rises in the air, and hear no sound.

A few of the golfers were pretty good. Two of them reached the green on which we stood on their first shot. Thus golf information arrived on the green in three waves. First, the light from the struck golfball announced the intended passage of the ball to the green. That information sped to my eye at the speed of light. Second, the sound of the struck ball arrived about a second later. Finally, a few seconds after that, the ball itself arrived.

Light waves, sound waves, or golfballs are all forms of information. They all share the quality of going from some "here" to another "here." To do that they must move, and can do so at various possible speeds. Could the information about the smacked golfball beat the speed of light? Could I have learned that the golfer hit the ball before the light reflected from the ball reached my eyes?

From looking at the table in the previous section of this chapter you saw that the fleaship somehow got stuck at $V = 1$, after being gone for quite a while from earth. Although the ship was continually being accelerated, it could never "cross over" the light barrier. This upper limit is a very important consequence of relativity. It sets a limit on the meaning of causality.

The Limits of Causality

Causality is the doctrine that states a relationship between a cause and an effect. It says that the event taken as the cause must occur before the event taken as the effect. The golfer hitting the ball is the causative event. The ball landing on the green is the effect. *Before* and *after,* however, are relative terms. How much before the effect-event should a cause-event occur in order to qualify as the cause of the effect? One second? A microsecond (a millionth of a second)? A nanosecond (a billionth of a second)?

Of course, it depends on how far apart in space the two events are. How far can they be separated? One light-second (the distance light travels in one second — 186,000 miles, or nearly two-thirds the distance to the moon)? One light-microsecond (982 feet, or nearly the length of a par-four golf hole)? One light-nanosecond (just under 1 foot)? Or perhaps even smaller?

Again, it depends. It depends on the speed at which the signal moves connecting the cause to the effect. And that is where the speed of light comes in. Since nothing can move faster than light, it appears that no prior event can be the cause of a latter event if the signal travels from the former to the latter faster than light.

But why not? The answer is that if such a signal could exist, it would be possible to view the prior and latter events simultaneously. In fact it would even be possible to view the events in the reversed order. To see this, consider the golf-course example again, only this time suppose that the oncoming party contains a golfer from outer space who has invented a magical "tachyonic" (faster than light) golfball. Imagine my surprise now. He hits the ball and it arrives on the green before I see him hit it! It is only after it appears on the green that the light signal from the golfer arrives and a second later the sound of the stroke. Suppose he could hit the ball so that it traveled at ten million times light-speed. Now suppose that the fairway to the course stretched along a highway. A motorist speeding along the road at 66.96 mph would see the ball being struck by the golfer at the same time the ball hit the green. If he decided to speed up to 70 mph, he would see the ball hit the green before it was struck. This time-reversed view could be had by any observer who happened to be passing by the golf course fast enough.

But why should the events reverse in time? How is it possible? It is possible because time sequences are not "nailed down." Looking back at Table 1 shows us how the time interval separating any two events is not fixed. The flea experiences life on the tennis-ball spaceship year after year. Each year lasts one year for the flea, but to those at home these "fleaship years" get longer and longer as the flea approaches light-speed. Between "fleayear" 0 and "fleayear" 1, earth ages 1.13995 years. If the flea left earth at 12:00 A.M., January 1, 1990, then on New Year's Day 1991 aboard the spaceship, earthlings would already be about fifty-one days and two hours into the new year. It would be February 20, 1991, and just about 2:00 A.M. at the space launch station.

On New Year's Day 1995 at midnight aboard the fleaship, it would be January 5, 2061, about forty-three seconds past midnight at the launch station. By the end of the flea's fourteenth year in space the earth would have aged 583,262 years, more than a half-million years.

We can look at this another way. From the earth's point of view, time aboard the fleaship slows down. The closer the fleaship approaches light-speed, the slower goes its time. The limit to all this is just at the speed of light. Here zero time for the flea is eternity for the earth. In other words, time does not pass for anything moving at the speed of light.

Taking our imagination even further, suppose that the flea leaves the tennis ball and reaches the speed of light (he really can't do this, because he has a tiny mass, but for the moment let this pass). Our flea becomes a photon (a particle of light). The flea's-eye view of the universe is eternally "now." At light-speed he would not suffer another fourteen years of existence. He would not experience time. All events that he is involved in would be simultaneous events.

Let us bring him back to earth for a moment. Picture the flea moving from the smacked golfball at the speed of light. From the moment the ball is smacked to the instant I see the flea on the green takes one microsecond. For the flea both events occur at the same time. Let the flea fly to the moon. It will take under two whole seconds for it to go from the fairway of earth to the astronaut's footprint. But the footprint and the fairway events are simultaneous for the flea. Let the flea fly as far as it will, taking as long as we wish, and the start and end of it all is simultaneous for the flea.

Understanding the "New Physics" of Spirit — The Quantum Wave Function

In the Beatles' song "Lucy in the Sky with Diamonds" we are asked to picture ourselves in a boat moving down a river. Alongside of us are trees growing tangerines and above us are skies filled and striated with marmalade colors and textures. We hear a seductive and sensuous voice calling us from the shore. We see the girl of our dreams and she has eyes so mysterious that they turn colors and form symmetries in myriads of kaleidoscopic visions.

Why not? I mean, why isn't the world filled with magic and mystery? Why does it seem so painfully ordinary? Indeed, why is it so painful at all? During the period of time starting with the mid-1960s I began to sense that visions of magic would again start to appear in the world. We were able to change the course of history by simply agreeing to change it. We ended our involvement in the Vietnam war.

I remember the discussions I had with other "wishful thinkers" who believed that the inevitability of war was nonsense. We said, and later heard it reverberate around the world, "All we have to do is declare ourselves the winner and leave." At the time that we spoke those words, all rational thought appeared against it. Our idea was declared nonsense. There were countless reasons why this idea could not and would not work.

But it did. And now we look back and find all kinds of reasons why we pulled out of Vietnam. We see the logic behind the decision. But when the "will" to pull out was strengthening, there were logical, compelling reasons preventing it. Our refusal to continue the war was, at the time we chose to not continue it, illogical and irrational. It was crazy. And it was wisdom. As we entered the seventies I felt heartened. I felt so heartened that the world could be turned into a "magical mystery tour" filled with plasticine people wearing mirrored, reflecting ties that I quit my job. I decided to be a real magician. Not

36

a conjurer, mind you, but a cosmic trickster, one who deals in the reality of this looking-glass world.

I had the perfect education for it: I was a professional physicist. All of the credentials were mine: bachelor of science in engineering physics, master of science in applied physics, and, through my ability proven by original research, doctor of philosophy in physics. I had reached the pinnacle of verifiability.

Physics is the study of the physical world. At its best it is a constant and never-ending search for the answer to the question "How is it that anything is?" In other words, at its best it is the search for how reality becomes existence.

Physics implies something in this search, namely that there are rules to the game. One rule is that there is something rather than nothing, that there is a reality that is capable of being experienced. Even if all that we sense is an illusion, there is something in back of that illusion that is real. This is the only rule that physicists practice when they are really doing physics.

At its worst, physics is redoing what has already been done. It is attempting to fit into context any data that do not fit into context. It is the desire to say to any phenomena that cannot be explained by the already existing theoretical structures of thought, "You are a hoax, an illusion." At its worst its skeptics become the leaders of frontier thinking. Put a skeptic at a frontier and nothing passes through the frontier gates into the homeland except those thoughts and ideas that fit the rules decided by the majority of the thinking populace. Frontiersmen become policemen, not hunters and seekers.

Breaking the bounds of mechanical thought is not easy unless you know how. To know how, you must be able to perceive how logic and rationalism, logical positivism, and mathematical and causalistic thinking are limited. Surprisingly, it is the fields of mathematics and physics that point to this limit. In other words, by going as far as we can go using rationalism as a base, which is the tool of science, we have come to see that the world that physics has discovered is not rational.

The discovery of the irrationality of the world came about with the realization of the quantum nature of all physical processes. Anything that involves matter involves the quantum. If it is, it is a quantum process.

The basis of quantum physics is the quantum wave function, or, as I choose to identify it, the qwiff. A qwiff is a nonphysical entity. It is a mathematical function. It is a form representing ideals, concepts that could, did, or would exist. It contains possibilities the same way an object contains attributes.

First we will look at how a physical attribute can be represented by a probability function. For example, a coin contains sides. When a coin lands after being flipped, it lands on one of its sides. It cannot land so that both sides are up. We use the coin as a "generator" of randomness utilizing its two-sidedness. We say that the coin has the probability factor of 50 percent heads – 50 percent tails built in. We also say that the coin's probabilities, P (heads) $= .50$ and P (tails) $= .50$, are mutually exclusive. The fact that the coin cannot be both heads and tails is the basis of a mathematical logic and a probability function for the coin.

We can summarize the logic and probability function of the coin by a simple table (Table 2).

Table 2

ATTRIBUTES	POSSIBILITIES		PROBABILITY BEFORE LOOK (PBL)		PROBABILITY AFTER LOOK (PAL)	
HEADS	TRUE	FALSE	.5	.5	1	0
TAILS	FALSE	TRUE	.5	.5	0	1
HEADS.AND.TAILS	FALSE	FALSE	0	0	0	0
HEADS.OR.TAILS	TRUE	TRUE	1	1	1	1

By "true" and "false" I simply mean that the coin has two possibilities for each attribute. If the coin has heads showing, it cannot appear with tails showing. So that in the column where heads is "true," tails must be "false." Similarly, if tails is "true," heads must be "false."

These choices are assigned numerical significance by the probability function. Now, in quantum physics probability functions are subject to rapid and discontinuous changes. These changes have physical consequences that, I believe, are the basis for consciousness. Thus it is important to realize that the PBL — the probability before looking — and PAL — the probability after looking — are not the same and that one changes to the other with the simple act of awareness.

Thus the probability of "true" before we actually look at the coin is 50 percent because there are two possibilities and one of the two is heads. The probability of "false" before we actually look at the coin is also 50 percent for the same reason. But after we look at the coin, these probability functions change dramatically.

Once we see the coin, we "know" the truth or falsity of the statement "The coin shows heads." Consequently the probabilities change after observation. They change discontinuously as we learn from our experience which is true and which is false.

As we can see, there is no way that the coin can "be" heads and tails at the same time. The statement that it is "heads and tails" is always false or always has the probability zero. Similarly, the statement that the coin is "heads or tails" is always true or always has the probability one. These statements are in themselves always true both before and after observation of the coin. That is, before looking at the coin it's true that the coin is heads or tails. Looking at the coin does not destroy the truth of the statement; it is just a fact that we know which side is showing. Similarly, before looking at the coin we know that it cannot have heads and tails showing at the same time. If it did, it wouldn't be a coin. After looking at the coin and realizing which side is showing, it is obvious that it cannot have both heads and tails showing at the same time.

In quantum physics these two simple statements are also taken for granted. They are called, respectively, *orthogonality* and *completeness*.

The .AND. statement, the one that says "HEADS.AND.TAILS," is a statement of orthogonality. The .OR. statement "HEADS.OR.TAILS" is a statement of completeness. (I have used the .AND. and .OR. to stand for "and" and "or," respectively. This is consistent with modern computer programming language. They mean exactly what is meant by these logical functions in computer usage.)

Orthogonality means mutual exclusivity, that there cannot be two things in the same place at the same time. Completeness means inclusivity, everything that is is included, and nothing is missing. Thus orthogonality means the existence of the impossible, and completeness means the existence of certainty. Orthogonality and completeness mean the existence of falsity and truth.

So far, so good. A coin in its reality appears no different from a quantum-physical object. If that was all there was to quantum physics, there would be no magic in it, just ordinary probability functions and nothing more.

But there is something more. There is a logical paradox built into quantum physics. As we will see, completeness also implies the ability to choose from complementary attributes. It is this complementarity that introduces the paradoxical nature of the world.

Quantum Paradoxes: The Common Thread That Pulls Us Out of Space-time and Matter

To understand this vital difference, I ask you to consider the simplest quantum-mechanical system I can think of, a particle in a tube. Imagine that a ball, a very tiny ball, has been placed into a tube, a very tiny tube. Rid yourself of any gravity or friction. This means that once we shake the tube, the ball moves and does not stop because of rubbing against the sides of the tube. Now close the tube tightly and shake it about a bit.

Now there are two basic things we can discover about the ball (or particle): It exists at some location inside the tube and it is moving with some energy inside the tube (including the distinct possibility that it is moving with zero energy, i.e., not moving at all).

To make life simpler and yet not too far from what is actually predicted according to quantum physics, suppose that the ball is either on the right side or on the left side of the tube. These two possibilities are equivalent to the sides of the coin mentioned earlier. We could construct a similar "truth" table, only use "right" and "left" instead of "heads" and "tails."

But there is an additional attribute that complicates things: The particle is moving with some energy in the tube. Making things conceptually easier, we can assume that it is moving with either low or high energy. This leads to four distinct situations:

- The ball is on the left and moving fast.
- The ball is on the left and moving slowly.
- The ball is on the right and moving fast.
- The ball is on the right and moving slowly.

(Again, to avoid misconceptions and obvious difficulties, we can ignore the possibility that the ball is in the middle of the tube or that the ball is standing still. This is equivalent to ignoring the possibility of the coin landing on its edge in the previous example.)

For brevity, we can let "L" mean position left; "R" position right; "F" moving fast; and "S" moving slowly. The four statements of possibilities are symbolically:

- L.AND.F
- L.AND.S
- R.AND.F
- R.AND.S

which is certainly easier to write and remember. Now suppose we construct a truth table for the above possibilities. Certainly we have included them all so we expect that the table will be orthogonal and complete. We would expect it to look like Table 3.

Table 3

ATTRIBUTES	POSSIBILITIES				PBL				PAL			
L.AND.F (LF, left/fast)	T	F	F	F	.25	.25	.25	.25	1	0	0	0
L.AND.S (LS, left/slow)	F	T	F	F	.25	.25	.25	.25	0	1	0	0
R.AND.F (RF, right/fast)	F	F	T	F	.25	.25	.25	.25	0	0	1	0
R.AND.S (RS, right/slow)	F	F	F	T	.25	.25	.25	.25	0	0	0	1
LF.AND.LS.AND.RF.AND.RS	F	F	F	F	0	0	0	0	0	0	0	0
LF.OR.LS.OR.RF.OR.RS	T	T	T	T	1	1	1	1	1	1	1	1

In this table T and F stand for true and false, respectively; PBL and PAL stand for probability before looking and probability after looking, respectively. Again, there are no surprises here. Given four possibilities, with each possibili-

40

ty equally likely, gives a probability of 25 percent for each shot before a "look-see" and the usual "collapse of the probability function" after seeing what takes place.

Thus, for example, the discovery of the truth of L.AND.S (in words, the particle is on the left and moving slowly) changes the probability of this from 25 percent, or one out of four, to certainty after the truth of this is discovered.

What is remarkable about this table is that for a real particle in a tube (real in the quantum mechanical world), this table does not describe reality at all. Instead we find Table 4, which is blank.

Table 4

ATTRIBUTES	POSSIBILITIES								PBL				PAL			
L.AND.F (LF)	F	F	F	F	0	0	0	0	0	0	0	0
L.AND.S (LS)	F	F	F	F	0	0	0	0	0	0	0	0
R.AND.F (RF)	F	F	F	F	0	0	0	0	0	0	0	0
R.AND.S (RS)	F	F	F	F	0	0	0	0	0	0	0	0
LF.AND.LS.AND.RF.AND.RS	F	F	F	F	0	0	0	0	0	0	0	0
LF.OR.LS.OR.RF.OR.RS	F	F	F	F	0	0	0	0	0	0	0	0

There is no probability for any of these attributes to appear. To see how ridiculous this is, let me add that it is perfectly possible for the particle to be on the left of the tube and be moving either fast or slowly, or L.AND.(F.OR.S) can be true and in fact is true 50 percent of the time. But the statements "The particle is on the left and is moving fast" and "The particle is on the left and is moving slowly," (L.AND.F).OR.(L.AND.S) can never be found true. To see this, consider Table 5 on page 42.

First of all, notice that certain rows are identical. This means that they say or mean the same things. For example, the statement "The particle is on the left of the tube" is the same as "The particle is on the left of the tube and it is moving either fast or slowly" (see Table 5A on page 42).

Next notice that there are asterisks (*) in certain columns. These asterisks are vital. They say that as far as we are concerned they are on a different planet. In other words, nothing can be said about the truth or falsity of an asterisked entry in a row of the table. For example, in Table 5A the statement that the particle is on the left of the tube, L, implies no knowledge at all about the way it is moving. That is why the statements shown in the two rows above are identical. On the other hand, if the particle is moving fast, we really say nothing about its location right or left in the tube. Thus, quantum mechanics

41

Table 5

ATTRIBUTES	POSSIBILITIES				PBL				PAL			
L	T	F	*	*	.5	.5	*	*	1	0	*	*
R	F	T	*	*	.5	.5	*	*	0	1	*	*
F	*	*	T	F	*	*	.5	.5	*	*	1	0
S	*	*	F	T	*	*	.5	.5	*	*	0	1
L.AND.F	F	F	F	F	0	0	0	0	0	0	0	0
L.AND.S	F	F	F	F	0	0	0	0	0	0	0	0
R.AND.F	F	F	F	F	0	0	0	0	0	0	0	0
R.AND.S	F	F	F	F	0	0	0	0	0	0	0	0
L.AND.(F.OR.S)	T	F	*	*	.5	.5	*	*	1	0	*	*
R.AND.(F.OR.S)	F	T	*	*	.5	.5	*	*	0	1	*	*
F.AND.(L.OR.R)	*	*	T	F	*	*	.5	.5	*	*	1	0
S.AND.(L.OR.R)	*	*	F	T	*	*	.5	.5	*	*	0	1
L.AND.R	F	F	*	*	0	0	*	*	0	0	*	*
L.OR.R	T	T	*	*	1	1	*	*	1	1	*	*
F.AND.S	*	*	F	F	*	*	0	0	*	*	0	0
F.OR.S	*	*	T	T	*	*	1	1	*	*	1	1

Table 5A (detail)

L	T	F	*	*	.5	.5	*	*	1	0	*	*
L.AND.(F.OR.S)	T	F	*	*	.5	.5	*	*	1	0	*	*

indicates that we cannot know simultaneously both the energy of motion and location of the particle.

But this "unknowing" of movement is realized as location. In other words, the location of the particle is complementary to energy of motion. This realization can occur as location on the left or location on the right. Similarly, the knowledge that the particle is moving fast is a realization of the "unknowing" of the particle's location. Simultaneous "knowing" of complementary attributes is not possible.

For example, comparing the entries for R (particle on the right) and F (particle moving fast), we see that these two experiences or attributes are "in different spaces." It's as if knowledge was somehow captured by one space or the other, but not both.

This is especially evident after inspection of the last four rows of the table (Table 5B). There are other relationships to be found in the table. What is most important is the realization that quantum physics denies us our ordinary "commonsense" notion about location and movement in space-time.

Table 5B (detail)

R			F	T	*	*	.5	.5	*	*	0	1	*	*
F			*	*	T	F.	*	*	.5	.5	*	*	1	0

It also introduces a new and startling possibility. These spaces are complementary to each other. We are given choices. These choices are not to be found, however, in the following entries in Table 5C.

Table 5C (detail)

L.AND.F			F	F	F	F	0	0	0	0	0	0	0	0
L.AND.S			F	F	F	F	0	0	0	0	0	0	0	0
R.AND.F			F	F	F	F	0	0	0	0	0	0	0	0
R.AND.S			F	F	F	F	0	0	0	0	0	0	0	0

No such observations are possible. All of these possibilities are ruled false. We, the observers of reality, bring into existence one form of reality or another. We do this by conscious choosing. Either we choose to observe the position of the particle in the tube or we choose to observe its energy of motion.

When we measure position (Table 5D, next page) we have no way to determine the energy of motion (Table 5E, next page).

By choosing to measure *M*, motion, we lose information about *P*, position, and vice versa. That is the true meaning of the asterisks. They symbolize our consciousness entering into the picture. By our choices we enter into one of two parallel worlds of choice, position or energy of motion.

43

Table 5D (detail)

L	T	F	*	*	.5	.5	*	*	1	0	*	*
R	F	T	*	*	.5	.5	*	*	0	1	*	*

Table 5E (detail)

F	*	*	T	F	*	*	.5	.5	*	*	1	0
S	*	*	F	T	*	*	.5	.5	*	*	0	1

The paradox of this is that we cannot do both. We cannot determine both even though our common sense tells us that we should be able to do both.

Now, in classical logic, the statements

- L.AND.(F.OR.S)
- (L.AND.F.).OR.(L.AND.S)

are entirely equivalent. Let me write them in English:

- The particle is on the left and it is either moving fast or it is moving slowly.
- The particle is on the left and it is moving fast or the particle is on the left and it is moving slowly.

While statement one can be either true or false, statement two can never be true according to quantum logic.

We can put both statements in Table 5F.

Table 5F (detail)

L.AND.(F.OR.S)	T	F	*	*	.5	.5	*	*	1	0	*	*
L.AND.F (LF)	F	F	F	F	0	0	0	0	0	0	0	0
L.AND.S (LS)	F	F	F	F	0	0	0	0	0	0	0	0
LF.OR.LS	F	F	F	F	0	0	0	0	0	0	0	0

Classical logic would have it that both the first and last rows of Table 5F would contain "truth" values. The quantum world necessitates otherwise.

Earlier, I stated that a perceived attribute is a created potential for its complementary attributes. For example, we saw that L meant the same thing as L.AND.(F.OR.S). The idea is that somehow the F and the S are "getting together," superposing themselves, conspiring to produce the L experience. Consequently, the "spaces" *P* and *M,* meaning the ideal possibilities of choosing to look at position or motion, are connected.

L is a way of seeing F.OR.S. R is another way of seeing F.OR.S. This means that there are two ways of combining the attributes of the *M* space to produce the attribute of the *P* space. Similarly, F is a way of seeing L.OR.R. And S is another way of seeing L.OR.R. Thus there are also two ways of combining the attributes of the *P* space to produce the attribute of the *M* space.

In other words, the experience of discovering the particle on the left of the tube equals a potential reality of the particle moving fast or slowly. This potential reality (F or S) is equivalent to the actual reality of finding the particle on the left of the tube. Reality is thus both real and paradoxical. It is real for what is observed and it is paradoxical for the complementary observables.

When we experience a reality we are disturbing the universe because we are creating a particular combination of potential realities. If we discover the particle on the right, we have created one particular combination of potential movements.

In this simple example it is quite easy to determine how these ideals should be mathematically represented. Movement, or *M* space, contains only two dimensions: S and F. Similarly, Position, or *P* space, contains only the two locations R and L. We can represent these spaces alongside each other in Figure 8.

COMPLEMENTARY SPACES OF THE MIND

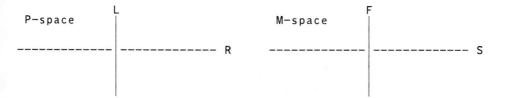

Figure 8

These spaces are called complementary Hilbert spaces. A single line in one space corresponds to a single line in the other. The "reality" of an experience in one of the spaces always means that the line is along the axis line symboliz-

45

ing that experience. For example, the experience L is shown in both spaces as Figure 9.

THE EXPERIENCE THAT THE PARTICLE IS ON THE LEFT SIDE OF THE BOX

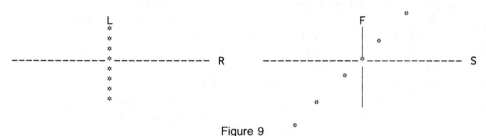

Figure 9

The asterisk line is placed along the L axis in *P* space. It is equivalent to the diagonal asterisk line across the upper-right and lower-left quadrants of the *M* space. This can be thought of as a combination or superposition of F plus S. Similarly the experience R appears in Figure 10.

THE EXPERIENCE THAT THE PARTICLE IS ON THE RIGHT SIDE OF THE BOX

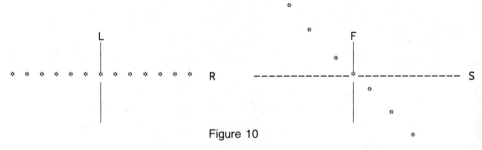

Figure 10

This can be thought of as the combination or superposition F minus S. In a completely analogous manner the experience F would appear as in Figure 11, while S would look like Figure 12.

Thus the location of the particle on the left side of the tube means the same thing as the particle is moving fast and moving slowly at the same time. In other words, F and S are added together. In the language of quantum physics this means that the quantum waves, or qwiffs, representing fast moving (high energy) and slow moving (low energy) are actually superposed; one is laid out upon the other and the numbers are added and adjusted or renormalized. This is called the principle of superposition.

THE EXPERIENCE THAT THE PARTICLE IS MOVING FAST (HIGH ENERGY)

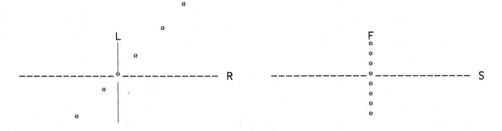

Figure 11

THE EXPERIENCE THAT THE PARTICLE IS MOVING SLOW (LOW ENERGY)

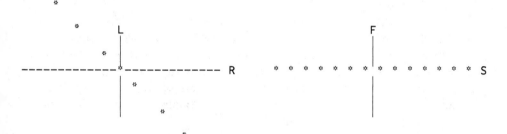

Figure 12

The processes of learning about the particle in the tube would produce a certain kind of movement in the complementary Hilbert spaces: The asterisk line would appear to rotate discontinuously according to what was being experienced. If, for example, I measured the particle to be moving slowly, S, and then measured the location of the particle, I would have no way of controlling which way the asterisk line would "spin" in the P space. If it snapped "vertically," my next measurement of P would be L. If it snapped "horizontally" in the P space, my next measurement of P would be R.

From the above, we could begin to conclude that space-time "space" is a limited concept for basing experience. Hilbert space, even in the simple example illustrated by the particle in the tube, shows a far richer "context" for basing experience. It is a dynamic, or "living," space, a "living room" if you wish.

47

The space-time context, however, is taken as our map of reality. We relate our experiences according to where, when, and how much. Quantum physics shows us that we really can't do this. It is an illusion that we can. The paradox is that we think, sometimes quite doggedly, that this illusion is real. The quantum is that embarrassing little piece of thread that always hangs from the sweater of space-time. Pull it and the whole thing unravels.

Quantum Oneness (Solipsism)

Although the qwiff has paradoxical features pointing to the possibility that space-time is not fundamental, it still appears to be a mechanical concept. Even though the location and the motion of an object cannot be determined at the same time, the measurement of an attribute is "composed" of a superposition of its complementary attributes. This has led many physicists to seek the "underlying hidden mechanics" that "caused" the transformation to occur from the attribute of position to its complementary attribute of motion, or vice versa, whenever a corresponding measurement was performed.

Perhaps that was all there was to it. There were hidden connections brought into play whenever an experimenter placed a motion- (i.e., energy-) or position-determining apparatus in the tube. The attempts to "grasp" the particle led to uncontrollable influences on this hidden variable. Consequently, with repeated attempts to measure position and then motion in the tube there would appear a quality of randomness in the distribution of results obtained: Sometimes the energy would be high with the particle on the left of the tube and about half the other times it would be low. A similar result would be found after locating the particle on the right side of the tube.

In fact, the possibility of hidden controlling parameters cannot be ruled out. It is possible to have a mechanical underpinning to the universe, including the paradoxical results discussed in the previous section. Several physicists, including David Bohm and John Stewart Bell, have successfully constructed hidden variable theories for a single object in an energy field, such as a particle in a tube, a particle on a spring, in an atom, in a magnetic field, etc. However, a theory is only as good as its predictions. Hidden variable theories are based on Newtonian mechanics. They claim that the hidden universe of variables gives back to each particle its "commonsense" attributes.

This would mean that the particle in the tube would "really" have both a position and a specific energy at the same time in the tube. The inability to discover this and the need for the qwiff to represent the particle are therefore due to our human inability to "catch hold" of the hidden variable. With more refined techniques these hidden controlling factors could be uncovered, bringing about a return to the classical logical position where, for example, L.AND.F (the particle is on the left and moving fast) could have a 25 percent chance of existing.

But this isn't the whole story yet. Quantum physics had a further surprise. This was first discovered by Albert Einstein, Boris Podolsky, and Nathan Rosen (EPR) in 1935 at Princeton. I discussed this noteworthy paper in my previous book, *Taking the Quantum Leap.* EPR asked the embarrassing question, "What would happen if two particles were allowed to interact and then separate?"

According to what we have discussed earlier, neither particle could have both a well-defined location and energy at the same time. But suppose that an observer measures the location of one of the particles, would his observation shed any light on the location of the other particle? EPR showed that it was possible to construct a qwiff that had the rather strange property that observation of one particle's location "instantaneously" gave information about the location of the other particle. Or similarly, observation of one particle's energy gave instantaneous knowledge of the energy of the other particle. Yet, without that first measurement the second particle's attribute (as well as the first's, of course) was unknown.

EPR were clearly bothered by the instantaneity of the knowledge. How could the information from one particle travel so quickly to the other? We have already seen, in the earlier sections of this chapter, that simultaneity lacks causality. Simultaneous events cannot be causally related. If they were related in any sense of the meaning of causality, the events would be in violation of Einstein's theory. Does the qwiff violate relativity?

Such an apparent causality-violating quantum wave function is called *correlated.* A correlated qwiff means that once an attribute for one of the systems has been measured, the corresponding attribute for the other previously interacting system is also determined. It also means that before any measurement is made, neither system possesses a unique attribute. Instead it is their combination, their lack of identity, that possesses an attributable value. For example, each particle may not have a unique energy, but both particles together do.

Before I give an example of this, I want to point out that in 1957 David Bohm and Yakir Aharonov discussed an experimental proof for the EPR "paradox," as it came to be known later. They showed that any interpretation of experiments involving two previously interacting particles where the qwiff describing the two particles did not have this correlation "built in" would fail. They then gave a working possibility for carrying out an experiment where two particles, each having an intrinsic "spin" (meaning that the particles can be thought of as each spinning on an axis running through them), separate after having interacted, and spin measurements performed on one instantly affect the spin of the other.

In 1964 John Bell showed that even though a hidden variable understructure was possible for the quantum-mechanical EPR paradox, it in itself would contain a relativity-violating communication link, traveling faster than light-speed between the particles. Since the hidden variables were created to rid us

49

of, among other things, such violations as these in the first place, it appeared that hidden variables were unnecessary and a further cumbersome detail. Quantum mechanics with its qwiffs was weird enough already.

Since the appearance of Bell's theorem many papers have been published showing how deeply ingrained this quantum correlation is. Hidden variable attempts still make the "physics scene." No one has yet been able to rid us of the "dreaded" faster-than-light — information-traveling — quantum-correlated connection that must exist between any two or more previously interacting systems.

Once having interacted, the correlation persists until another system outside the correlated system interacts with it. This interaction with the outside system alters the correlation irrevocably. The result of this could be a memory trace if the outside system is consciousness itself. We will examine this hypothesis later.

Looking again at the original correlation, this link between two or more parts of the correlated system means or implies a unity — a special unity to be sure, but a wholeness nevertheless. All it takes to create this unity is interaction. Both parts of the interacting system leave the interaction region with more than they came in with (they are now part of a greater whole) and with less than they came in with (they each have suffered a loss of an individual attribute).

Now it is time to look at another simple (I hope) system that possesses this amazing link. We will look at a tube containing two particles and we will suppose that the two particles have equal masses but are distinguishable in some other attribute (they might have different "spins" or, if they are molecules, different molecular arrangements such as found in sugars). Because the particles possess masses one cannot be in the same location as the other at the same time — at least we would hope not. This is their interaction. Whenever they "touch," their mutual quantum wave has a "node." This means that whenever and wherever particle one's location is the same as particle two's, the quantum wave must be zero. In other words, we are saying that it is not possible for both particles to be in the same place in the tube at the same time. There are no odds for this to happen. The probability function is zero — it is impossible for it to occur.

Without this "touchiness" between the particles there would be no restriction. Both would be free to roam the inside of the tube, free, that is, except for the obvious boundary of the tube's walls. These boundaries include the walls at the ends of the tubes as well. Again the particles cannot penetrate the walls. This means that the qwiff also has wall "nodes." It vanishes if either particle touches any wall in the tube.

Now it is important to realize that the "wholeness" of quantum physics arises in the qwiff. For a single particle in the tube the concept of wholeness is unnecessary. But with two particles present we do not have two separate qwiffs representing them; they are represented by a single qwiff.

It is also necessary to realize the effect of boundaries on quantum systems. The tube's walls (both the tubular and end walls) are "boundary conditions." They act as limits. They provide the basis where completeness and orthogonality become possible. Without these limits — that is, without boundary conditions — such facts as repeatability of experience and understanding of facts are not possible. In other words, the ability to repeat, memorize, deduce — in general, think — depends on the choice of boundary conditions, the limits we set for our consciousness.

Problems in language translation and human communications may generally be due to a lack of commonly applied limits to meanings. By failing to limit the meaning of words, miscommunication is bound to occur. I believe that this failure is quantum physical and caused by unrecognized attempts to repeat situations with different boundary conditions being utilized.

In the example of the particle in the tube, the walls restrict the unlimited qwiff; they hold it down, making it vanish at their surfaces. This means that the probability of finding the particle there is zero. And it is this limitation of freedom that makes the different energy states, fast and slow, orthogonal to each other. The boundary conditions determine the allowable energies. Any energy is not possible; there are only certain energies that the particle can "enjoy." These energies are separated from each other in the same sense that whole numbers are separated from one another. In fact it is this quality of the appearance of orthogonality and separate countability that gives quantum mechanics its quantumlike character (the word *quantum* means a whole amount such as one or two, but nothing in between). A quantum jump is nothing more than a sudden change from one energy level to another. By changing the boundary conditions we change the probabilities and the energies involved in these jumps.

In a similar way, I believe, the words we choose to speak or think limit our energies by altering the physical boundary conditions of our human nervous systems. By changing those boundary conditions we may be able to change the energetic possibilities we have at our disposal. This may explain the problem of human metabolism: Why, in the case of two equally active people eating the same food, one person gains weight while the other loses or maintains weight. In the first person's case there are energy transitions taking place in which molecular structures comprising tissue or fat storage are created. In the second person's case these "fat-storage energy levels" differ from the first person's because the second person's thought processes alter the boundary conditions necessary for molecular fatty tissues to be built up. Consequently transitions to these levels take place with a reduced probability. The energy goes into other processes such as enhanced sensing or thinking.

Indeed, the exciting possibility exists that through altering our thinking processes, and thereby our mental boundary conditions, we may be able to make ourselves smarter by altering the number of energetic brain transitions — thereby literally providing more food for our brains. This modification, while useful in the land of plenty, could serve little use in a poor land where body weight becomes necessary for primitive manual labor.

Just as the boundary conditions determine the energy-level structure within the quantum system, they also determine the availability of these energy states to provide complementary attributes. We could say that the "quality" or accuracy of determination of the location of the particle in the tube depends on the shape of the qwiffs corresponding to the different energy states. These shapes, in turn, depend on the boundary conditions. We will say more about this later.

Returning to our "simple" example (nothing is simple in quantumland), with both particles interacting, thus introducing the further limitation that the qwiff must vanish whenever and wherever the particles find themselves touching, it becomes impossible for both particles to enter the same energy state. They both cannot be in the slow-moving or in the fast-moving lane. This restriction is extremely important. It enters into every atom in existence that contains more than a single electron. We will return to this exclusion principle in Chapter 6.

If one particle is slow, the other must be fast. However, the exclusion introduced by the "untouchability" factor takes away the certainty of knowing which is which. It must be equally possible to find particle one fast and particle two slow, or particle one slow and particle two fast. Thus the system of two particles is correlated through the qwiff. Both particles together have energy, but neither particle separately has a defined energy with any certainty.

To display the qwiff for two or more particles presents something of a problem. The reason is that each particle really needs three dimensions in order to be shown. Since it is possible to imagine holding one particle still while moving the other, a two-particle qwiff requires six dimensions to be seen. By suppressing two dimensions for each particle we can show a two-particle qwiff. In order to do this we need to display the one-particle qwiff as a vertical column, as shown in Figure 13.

This shows a particle on the left side of the tube. The only difference is that "left" is up in the picture and "right" is down. Similarly, it can be displayed horizontally as also shown in Figure 13. We show this in order to picture the new situation, namely, when two particles are placed in the tube. We need to separate the location of one of the particles from the location of the other. This is easy to do in "real" space but somewhat awkward in the qwiff space we are showing.

To accomplish this we let the vertical direction show the location of particle one and the horizontal direction locate particle two. Our tube will then appear like a crossword puzzle, where each and every number says something about *both* particles.

Figure 14 (on page 54) shows the qwiff for two noninteracting particles in a tube. The vertical distance shown and the horizontal distance shown correspond to only one length of the tube. The display is as shown in order

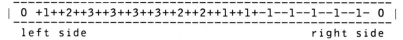

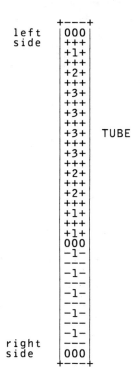

Figure 13

to present a view that includes both particles. Thus by moving your eye horizontally along a row you are looking at what happens to the qwiff when the location of particle one is held fixed and the position of particle two is varied. Similarly, by moving your eye vertically up or down a column you see how the qwiff changes when the location of particle two is fixed and the position of particle one is varied.

The upper-left-hand corner of the diagram marks the left side of the tube for one particle and the left side of the tube for the other. The lower-left-hand corner marks the left-hand side of the tube for one particle and the right-hand side for the other, etc. Thus if we look at a particular spot, say, the spot marked *2* in Figure 14, it shows the qwiff for particle one five units from the right side of the tube and particle two three units from the left side of the tube. Here we see that both particles tend to be in the center of the tube and both particles have low energy. Although there appears to be a greater tendency to find the particles "centered," we really can't tell whether either particle is on the left or right side of the tube. That's because of the highly symmetric pattern shown.

TWO NONINTERACTING PARTICLES IN A TUBE, EACH WITH LOW ENERGY

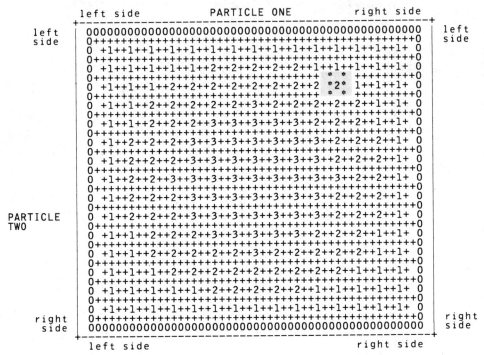

Figure 14

There is equal tendency, i.e., equal probability, of locating either particle on either side of the tube.

In Figure 15 both particles have high energies. The plus and minus signs correspond to positive and negative values for the qwiff. We see the clear pattern of zeros forming a cross in the diagram. These zeros remain as fixed nodes as the particle's qwiffs vibrate in the tube. Again we are unable to say where each particle is to be found. We do know for sure that neither particle "occupies" the tube's center.

In Figure 16 (on page 56) we see the qwiff pattern corresponding to the two particles, with each having a more or less well-defined position but not well-defined energies. Each particle's qwiff would "slosh" back and forth through the tube as if the other wasn't present. In Figure 16 both particles tend to be on the left side of the tube.

But the above situations are not stable. We have ignored the interaction between the two particles. This interaction will correlate the locations of the particles or the energies of the particles, depending on what question we ask — i.e., whether we seek to measure one of the particle's energy or its position.

TWO NONINTERACTING PARTICLES IN A TUBE, EACH WITH HIGH ENERGY

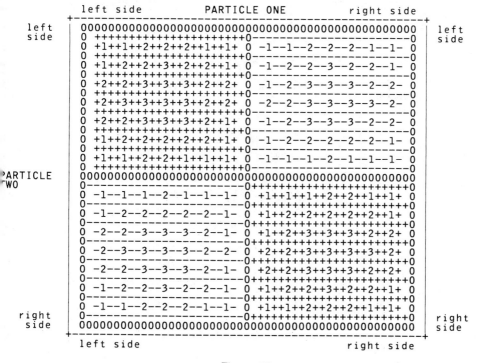

Figure 15

Allowing the particles "their space" in the tube, including their "touchiness," introduces a diagonal line of nodes across the diagram. In Figure 17 (on page 57) neither particle has a well-defined location or energy, but both particles together have a well-defined energy. In this paradoxical situation we see the quantum physics of solipsism, the physics of "wholeness."

Because of the correlation between the particles, the qwiff has a symmetry. This symmetry is called *antisymmetric* because the plus and minus signs are located as shown. For example, by scanning your eye along the fifth row until you reach the nodal "0" point and then upwards along the fifth column until you reach the top line, you see that the number pattern is the same except for a sign change. This holds true for every like row and column and means that if the two particles in the tube were to change positions instantly, the qwiff would automatically change its sign. The physical significance of this is that the two particles cannot both occupy the same position at the same time without their qwiff vanishing. The points where a zero exist are such points.

55

TWO NONINTERACTING PARTICLES IN A TUBE, EACH ON THE LEFT SIDE

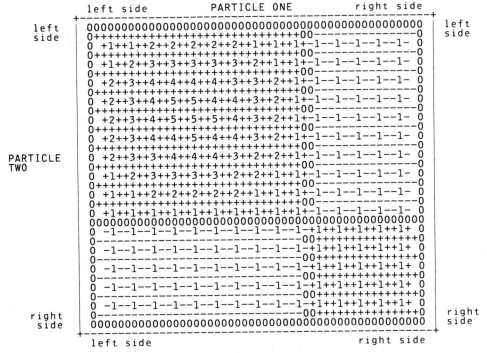

Figure 16

The values (from zero to 4) correspond to the strength of the qwiff "field." A 4 is a high value corresponding to a peak in the probability pattern. Thus we see that the qwiff is peaked positively and negatively ($+4+$ and $-4-$). Since the probability is computed by squaring the value of the qwiff, i.e., multiplying the qwiff by itself, both points have high probabilities of occurring. This means that the particles are likely to be discovered at the positions where the 4's occur. In the above this is at $+3/16$ and $-3/16$ from the center of the tube.

Figure 17 is composed of two "pictures," only this time the situation is "otherworldly." There are two complementary sets of picture pairs that can be superimposed to "create" the correlated stable picture. We can "add" together energy pictures or position pictures. Or, in other words, this illustration can be seen as composite of either two energy pictures or two positional pictures.

The reason this is significant is that what we do depends on our thoughts, i.e., how we think about the particles in the tube. For it is our thoughts that will lead to our disrupting the correlation. By disrupting it energetically we

TWO PARTICLES IN A TUBE INTERACTING AND CORRELATED

```
                left side        PARTICLE ONE       right side
              +--------------------------------------------------+
   left       |000000000000000000000000000000000000000000000000  |  left
   side       |000------------------------------------------0    |  side
              |0   0 -1--1--1--1--1--1--1--1--1--1--1--1--1- 0    |
              |0++000+--------------------------------------0     |
              |0  +1+ 0 -1--1--1--2--2--2--2--2--2--2--2--1- 0    |
              |0 ++++000------------------------------------0     |
              |0  +1++1+ 0 -1--1--1--2--2--3--3--3--3--3--2--1- 0 |
              |0+++++++++000--------------------------------0     |
              |0  +1++1++1++0 -1--1--2--3--3--3--3--3--3--2--1- 0 |
              |0+++++++++++++000----------------------------0     |
              |0  +1++1++1++1++1+ 0 -1--2--2--3--3--4--3--3--2--1- 0 |
              |0+++++++++++++++++++000----------------------0     |
              |0  +1++1++1++1++1++1+ 0 -1--2--3--3--3--3--3--2--1- 0 |
              |0+++++++++++++++++++++++000------------------0     |
   PARTICLE   |0  +1++2++2++2++1+ 0 -1--2--3--3--3--3--2--1- 0    |
   TWO        |0+++++++++++++++++++++++++++000--------------0     |
              |0  +1++2++2++3++2++2++1+ 0 -1--2--2--3--2--2--1- 0 |
              |0++++++++++++++++++++++++++++++000-----------0     |
              |0  +1++2++3++3++3++3++2++1+ 0 -1--2--2--2--2--1- 0 |
              |0+++++++++++++++++++++++++++++++++000--------0     |
              |0  +1++2++3++3++3++3++3++2++1+ 0 -1--1--1--1- 0    |
              |0+++++++++++++++++++++++++++++++++++++000----0     |
              |0  +1++2++3++3++4++3++3++2++2++1+ 0 -1--1--1--1- 0 |
              |0++++++++++++++++++++++++++++++++++++000-----0     |
              |0  +1++2++3++3++3++3++3++3++2++1++1+ 0 -1--1--1- 0 |
              |0++++++++++++++++++++++++++++++++++++++++000-------0|
              |0  +1++2++3++3++3++3++3++2++1++1++1+ 0 -1--1- 0    |
              |0++++++++++++++++++++++++++++++++++++++++++000-----0|
              |0  +1++2++2++2++2++2++2++2++1++1++1++1+ 0 -1- 0    |
              |0++++++++++++++++++++++++++++++++++++++++++++000---0|
              |0  +1++1++1++1++1++1++1++1++1++1++1++1++1+ 0  0    |
   right      |0+++++++++++++++++++++++++++++++++++++++++++++++0000|  right
   side       |000000000000000000000000000000000000000000000000  |  side
              +--------------------------------------------------+
                left side                         right side
```

Figure 17

will create the unstable situation where one particle has high energy and the other low energy (we won't be able to predict which has which). Or, by disrupting it "positionally" with a location-measuring device, we will create one particle on the left side of the tube and the other on the right. Again we won't be able to predict which is on which side. And that is the "whole" truth. Even though we only disturb one of the particles, the other immediately takes on a corresponding matching attribute. This happens instantly, faster than light can cross the distance separating them.

If we detect that one particle has low energy, we immediately "create" the other with high energy. Thus the pattern shown in Figure 17 discontinuously jumps to Figure 18 (on next page) as soon as particle one is found with high energy.

In Figure 18 particle one has high energy while the other has low energy. This is clear from the pattern of zeros running down the middle of the tube showing nodes. The vibrational pattern running horizontally has the higher energy while the pattern running vertically has the lower energy.

But another possibility must coexist along with Figure 18 in order to have

TWO NONINTERACTING PARTICLES IN A TUBE, ONE WITH HIGH ENERGY, ONE WITH LOW ENERGY

```
                left side        PARTICLE ONE        right side
        +------------------------------------------------------+
  left  | 0000000000000000000000000000000000000000000000000000 | left
  side  | 0++++++++++++++++++++++++0--------------------------0 | side
        | 0  +1++1++1++1++1++1++1+ 0 -1--1--1--1--1--1--1- 0
        | 0++++++++++++++++++++++++0--------------------------0
        | 0  +1++1++2++2++2++1++1+ 0 -1--1--2--2--2--1--1- 0
   L    | 0++++++++++++++++++++++++0--------------------------0
   O    | 0  +1++2++2++2++2++2++1+ 0 -1--2--2--2--2--2--1- 0
        | 0++++++++++++++++++++++++0--------------------------0
   W    | 0  +1++2++2++3++2++2++1+ 0 -1--2--2--3--2--2--1- 0
        | 0++++++++++++++++++++++++0--------------------------0
        | 0  +1++2++3++3++3++2++1+ 0 -1--2--3--3--3--2--1- 0
PARTICLE| 0++++++++++++++++++++++++0--------------------------0
 TWO    | 0  +2++2++3++3++3++2++1+ 0 -1--2--3--3--3--2--2- 0
        | 0++++++++++++++++++++++++0--------------------------0
        | 0  +2++2++3++3++3++2++2+ 0 -2--2--3--3--3--2--2- 0
   E    | 0++++++++++++++++++++++++0--------------------------0
        | 0  +2++3++3++3++3++2++2+ 0 -2--2--3--3--3--3--2- 0
   N    | 0++++++++++++++++++++++++0--------------------------0
        | 0  +2++2++3++3++3++2++2+ 0 -2--2--3--3--3--2--2- 0
   E    | 0++++++++++++++++++++++++0--------------------------0
        | 0  +2++2++3++3++3++2++1+ 0 -1--2--3--3--3--2--2- 0
   R    | 0++++++++++++++++++++++++0--------------------------0
        | 0  +1++2++3++3++3++2++1+ 0 -1--2--3--3--3--2--1- 0
   G    | 0++++++++++++++++++++++++0--------------------------0
        | 0  +1++2++2++3++2++2++1+ 0 -1--2--2--3--2--2--1- 0
   Y    | 0++++++++++++++++++++++++0--------------------------0
        | 0  +1++2++2++2++2++2++1+ 0 -1--2--2--2--2--2--1- 0
        | 0++++++++++++++++++++++++0--------------------------0
        | 0  +1++1++2++2++2++1++1+ 0 -1--1--2--2--2--1--1- 0
        | 0++++++++++++++++++++++++0--------------------------0
        | 0  +1++1++1++1++1++1++1+ 0 -1--1--1--1--1--1--1- 0
  right | 0++++++++++++++++++++++++0--------------------------0 | right
  side  | 0000000000000000000000000000000000000000000000000000 | side
        +------------------------------------------------------+
                left side                        right side

            H I G H     E N E R G Y
```

Figure 18

the stable energy situation shown in Figure 17. In Figure 19 (on page 59) we see this possibility.

By observing particle two with high energy, particle one immediately "snaps to" a lethargic state. It's as if two troopers were standing in line and when the command of attention is given, one trooper immediately slumps at the instant the other stiffens his backbone.

If instead of energy we attempt to disrupt the pattern by looking for one of the particles, i.e., measuring its position in space, the pattern will be disrupted in an entirely different way. In Figure 20 (on page 60) we see the qwiff

TWO NONINTERACTING PARTICLES IN A TUBE. PARTICLE ONE HAS LOW
ENERGY, PARTICLE TWO HIGH ENERGY.

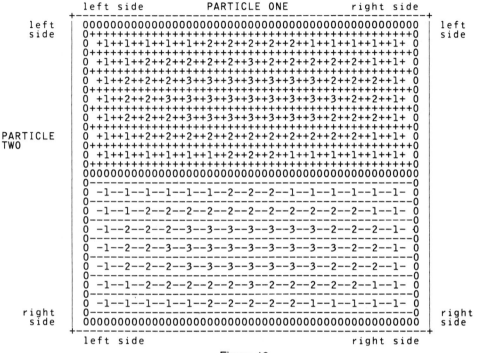

Figure 19

pattern corresponding to the two particles, each having a more or less well-defined position but not well-defined energies.

In this case we do know which particle is where. The "horizontal" particle is on the right side of the tube while the "vertical" particle is on the left side. This corresponds to the "peaking" of the qwiff in the upper-right quadrant of the diagram.

Again, another possibility must coexist along with Figure 20 in order to have the stable energy situation shown in Figure 17. In Figure 21 (on page 61) we see this possibility. When we observe particle two on the right side, particle one immediately appears on the left side, creating the exact reverse positions of those shown in Figure 20.

Thus we see in this "simple" example the meaning of complementarity: It depends on you; what you choose to look at drastically affects the qwiff patterns. I believe that our human brains operate in this manner. Each of our thoughts disrupts stable correlated patterns, momentarily producing the possibility of a new pattern. When we stop thinking, a natural correlation takes

TWO NONINTERACTING PARTICLES IN A TUBE, ONE ON THE RIGHT
SIDE, THE OTHER ON THE LEFT SIDE.

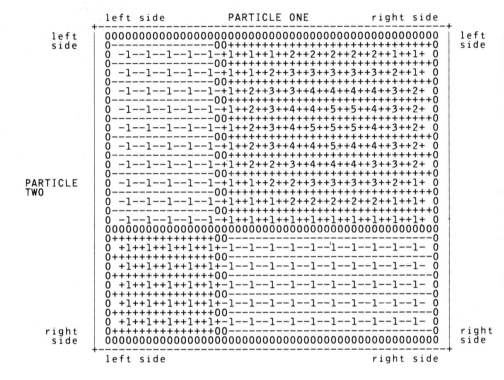

Figure 20

place through the actions of the interconnected neuronal nets that make up
our brains. The richer the nets, the higher the complexity, the greater the
possibility for new patterns forming. Perhaps when we sleep and turn off our
normal, survival-dependent consciousness, correlates can form that allow altered awareness states to occur. We will examine this again in Part Four.

Summary

There are two important concepts described here:

- Observation creates physical reality from idealizations. These
  idealizations are represented mathematically as qwiffs.

TWO NONINTERACTING PARTICLES IN A TUBE. PARTICLE ONE IS ON THE LEFT SIDE, PARTICLE TWO ON THE RIGHT SIDE.

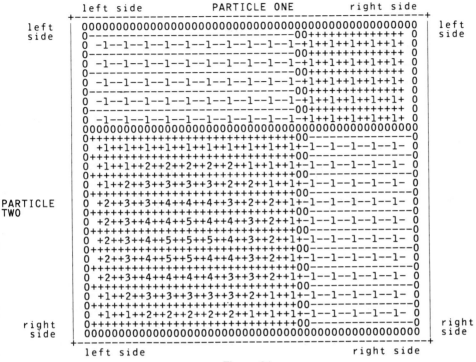

Figure 21

- Any observed reality is equivalent to a combination of two or more complementary idealizations. This combination is represented mathematically as a sum of qwiffs.

These ideas will be extremely important for understanding the role of consciousness, how learning and memory take place, and even how evolution can occur.

2

The Quantum Bridge
Between Mind and Matter

One extreme is the idea of an objective world, pursuing its regular course in space and time, independently of any kind of observing subject; this has been the guiding image from modern science. At the other extreme is the idea of a subject, mystically experiencing the unity of the world and no longer confronted by an object or by any objective world; this has been the guiding image of Asian mysticism. Our thinking moves somewhere in the middle, between these two limiting conceptions; we should maintain the tension resulting from these opposites.

Werner Heisenberg
Across the Frontier

Why Mind and Matter Cannot Truly Be Separated

In 1927 Werner Heisenberg announced a principle so profound and disturbing that it literally dislodged physics from its mechanical moorings. It was called the *principle of uncertainty* and it set definite limits on what could be known and, indeed, what was knowable.

The idea of uncertainty is disturbing enough as it is. Doubts plague us. Our relationships with others are at best surprising and, perhaps most often, unpredictable. We all feel keenly at times that we are victims of fate's cruelest tricks. Our only savior in all of this is the faith that underlying all of nature's mockery is an order, a smooth and absolute underpinning without which there would be no hope.

Although Heisenberg's pronouncement dashed this hope on the rocks of despair, many nevertheless held fast to it. Even today some physicists still believe that the principle of uncertainty, or indeterminism, is at worst a statement of ineptitude, of the inability of humans to master their physical environments. And even discussing Heisenberg's original proposal of the inability to determine simultaneously the position and the momentum of a subatomic particle, using a microscope that shines X rays on it, fails to illustrate the magic of his vision.

The reason for this has to do with the subtle distinction between an interaction and an observation. When two objects interact, they each experience a predictable alteration in their respective positions and/or momenta. In other words, they suffer predictable changes. Each object leaves the region of mutual interaction in a predetermined manner, the predetermination having been set by the interaction and the initial or start-up conditions of the objects.

Thus, for example, when an electron scatters from a collision with a proton in a typical high-energy event, its position in space becomes dependent on its history: How fast did it move? How close did it come to a dead-center hit? Also, the proton's behavior is dependent on the collision. Although quantum mechanics does not allow us to determine the exact location of the electron or proton, it specifies exactly how the particles influence each other in a

63

mathematical and lawful manner. Furthermore, their interaction provides a causal relationship between them. In no sense can we say that as a result of their interaction the proton "knows" the whereabouts of the electron. Each particle suffers inertial changes. No particle "knows" what it has suffered. We could say that the situation is "automated," even though we don't know where any particle is for sure.

On the other hand, an observation is unpredictable for a different reason. Any observation brings on a sudden and disruptive change, one that cannot satisfy a causal relationship. This sudden disruption was shown in the previous chapter in Figures 17 and 19. If we imagine, for the moment, that particle one possesses a primitive form of intelligence and take particle two to be "inert" — particle one is the "observer" and particle two, the "observed" — then Figure 17 shows the two in a correlated interaction. Their mutual qwiff displays their interdependence by the diagonal line of zeros slashing across the figure. Intelligence is not displayed here because the behavior of one particle is so connected to its partner. Furthermore the movement of the particles cannot be said to have independent energies. Here the intelligence of particle one is wasted. It is in an "automatic pilot mode."

But now suppose particle one "sees" particle two. Instantly the two particles are no longer correlated, no longer automated with each slavishly following the other. Each particle is suddenly "on its own." Each possesses an independent energy as shown in Figure 19. Here particle one is in a low-energy state while particle two is in a high-energy state. The act of observation has disrupted the universe. This act cannot be shown in either figure as a mathematical relationship. Rather, the sudden change from one picture to the other is the observational act. As a result, the two are actually no longer interacting. One "knows" about the other, but at a price. One is independent of the other.

It is here that Heisenberg's vision really belongs. When he says that one cannot observe both the position and momentum of any object at the same time, he really means "observe." He doesn't mean that one cannot interact with an object such that both its position and momentum are determined simultaneously. Albeit, this is a subtle distinction. It is the act of observing that disrupts the pattern so severely. Interaction does not cause such an acausal effect. It is always "causality-preserving."

When an observation of momentum occurs, the qwiff cannot support a correlated position for the objects. Their qwiff is "popped," positionally. Any and all positions are simultaneously present at the instant the momentum is "known." What we are dealing with here is the connection or correlation between concepts, concepts that are called complementary. As a result of an interaction between the two objects, their positions and momenta become correlated. Correspondingly, for each value of position there exists a range of values of momenta for each particle. Before an observation occurs, the ranges overlap, i.e., there is a mapping of the position of one particle onto the position of the other and, at the same time, a mapping of the momentum of one particle onto the momentum of the other. Even though we do not know the positions of either particle, nor their respective momenta, one undivided qwiff connects them, one undivided functional relationship that says that if you know something about particle one, you automatically know the same attribute about particle two.

But once an observation occurs, that precious connection vanishes. Knowledge of the position of one particle says nothing about the other particle's position from the time of the observation. In our example, particle one knows where particle two was at the instant particle one observed particle two. But in that act of observation, "one's" position is no longer correlated with that of "two."

Thus knowledge has a price, and that price is unpredictability. By simply and automatically following a causal, mechanical path, the observer cannot determine the future of objects that have been observed — especially if the path followed is the path of the observer himself. Mindless interaction, on the other hand, does allow such automated behavior to occur. By not knowing, particle one may continue its correlated "sense" of particle two's complementary attributes. However, particle one will not be in a position to "do" anything about it. Instead, particle one will be as compelled in its movements as is particle two.

It's as if the two particles were dancers mindlessly dancing on opposite sides of a gigantic stage, each mimicking the other's exact choreographed movements and yet each moving spontaneously. Then, with the sudden recognition that there is another dancer "there," across the stage, each dancer suddenly develops independent movements. Each has become conscious of the other and, at the same time, self-conscious. Now they must work to maintain their coordinated rhythms, the delicate nuances of each precisely copied movement.

Their dance reminds me of the ballet of Adam and Eve. While innocent and unmindful of each other, their movements in the sight of God were beautiful and symmetric. But then, with the inevitable bite of the apple and the serpent's way to the Tree of Knowledge, our happy, blissfully innocent couple became aware. In their awareness was a new freedom and responsibility. No longer were they automatons of God. Now they were forced to work and become creators just like their Creator before them. Godhood disturbs the universe.

Here, perhaps for the first time in human history, the role of mind in the universe of matter becomes apparent. Heisenberg's principle spells it out. To create is the master plan of the universe. Creation exacts a price for its freedom. It says that I must be free of the past correlations and interactions that bind me. "Untether me, let me loose of causality moorings. Do this, O Universe, and I shall work wonders for you. I will build and create machines, art forms, and people — all kinds of people and other living intelligences."

65

Thus it is that mind and matter cannot be truly separated. Mind is the outcry of indeterminism. It is the hope of matter. The function of matter is to interact with, and thereby correlate (thus building new structures), the universe — indeed build the universe itself. The function of mind is to tear down those very same structures, to analyze and decode nature's secrets, to inspect and create or recreate new structures. The universe is to be created. Mind is the creator.

How Our Observations Shape the Physical World

If mind is the observer and the creator, how does it function in the physical world? As far as quantum physics points out, observers have choices. These choices appear in the mind of the observer usually well before any observation actually takes place. The creative and weird element in this is that our normal perception of the world "out there" seems to tell us that we really have no choices at all. Our choices, no matter how logical and meaningful they may appear, are among sets of what physicists call complementary variables.

These variables literally have magical and far-reaching consequences, actually altering past, future, and present moments of our lives. Yet because of the illusory normalcy of these observations (they only appear mysterious in their capriciousness and in our inability to control them) we appear to be controlled by them. If God jokes with us, it is certainly through this paradoxical nature of the complementary variables of human observation.

To grasp the quantum nature of the physical world, physicists had to grapple with a new form of mathematical expression. This form was invented by Paul Adrian Maurice Dirac, a Nobel Prize winner in physics for his contributions to the theoretical understanding of quantum mechanics. (Later we will learn more about Dirac's *bra* and *kets*, his amusing names for abstract mathematical forms used in quantum physics, and how they conspire to produce our expectations in the physical world.)

Here we will point out that even though quantum physics leaves much to be desired in the way of causally being able to determine the results of our endeavors, it is by no means an impossible, no-win situation. Instead of each mathematical term on the physicist's writing pad representing a single observable consequence in the physical world (which was the theoretical understanding in Newtonian, or classical, mechanics), now he or she must deal with probability functions that, on paper at least, represent waves of uncertainty spreading out in space and through time.

These probability wave functions, or qwiffs (quantum wave functions), can be and are *under our control.* Along with qwiffs, we have in the theoretical world of quantum mechanics observables represented by mathematical symbols called "operators." Observables are the consequences of our actions. We "do" to observe. We must bring out or cause something to occur in order that we observe anything at all. To even read these words your eyes must do a dance

66

of incessant scanning. If they rest for the smallest instant, the image of the words will vanish from your retinas. Observables are the result of operations.

Operators operate on qwiffs. They are not qwiff "poppers," however.[†] Instead, they are qwiff "shapers." They shape and alter our expectations.

It is through the operators representing observables that minds gain control, as much control as is creatively possible, over our physical environments. At first an observable is no more than a wish, a dream. It is like a special net cast by a fisherman with a particular kind of fish shape designed into its mesh. He throws it out and hopes for the best. He cannot determine ahead of time exactly what he will catch. His fish-shaped mesh will allow some fish to escape and will capture some that are not the fish he desires.

This is the case when he first casts his net in a strange patch of ocean. In a similar manner an operator-observable acts on a qwiff. It shapes the qwiff so that any one of the many possible values attributable to the observable can be realized. To each observable are consigned a family of possible values. These values are observable because they are measurable, that is, they are counted and represented by numbers. All the operator does is set up the possibilities. Now the observer himself enters. He looks for the desired observable by "casting his net," i.e., setting up his instruments and measuring devices so that they will produce a value corresponding to his expectations. He cannot predict which fish he will catch. He has, so to speak, set his mind on a certain track. He is seeking from the qwiffian "ocean" a certain class of "fish."

Now he waits until his instruments indicate that he has made a "catch." Lights go on, dials light up, bells ring, and computer tape pours out of recording instruments. Just at that instant, he knows which particular value he has "caught." And just at that instant, that precise microsecond, the qwiff has popped. It now conforms to the will of its master. Its shape in space-time not only fits the mold of the observable but it conforms to the precise value it has just produced. It wants to "please its master" by reproducing the observed value once again and then again, as long as the observer desires. In this manner, i.e., by continuing to see the world the same, the world "is" the same. It appears secure and sensible. The data of our senses repeat and appear to our minds as lawful and causally meaningful.

Thus it is that operator-observables play a dual role in the universe. The first time the observer uses one, it "sets up" the physical conditions that will enable him to continue observing the same value for the desired observable.

---

†A qwiff "pop" is a sudden and unalterable, nonmathematical change in a qwiff, or quantum wave function. It occurs instantaneously and, once having changed in this manner, the qwiff cannot reverse itself by any mathematical causal transformation. That's why it is nonmathematical: No mathematical procedure can reproduce the observed result. When a qwiff pops, a particular value for a physical attribute and the knowledge of that "expectation" value of that attribute are simultaneously created in the mind of an observer.

But there is a "Catch-22" to this. Not all observables are repeatable. Their operators must follow another quantum rule if they are to reproduce identical values on repeated attempts at measurement. This rule was first discovered by Werner Heisenberg before he announced his uncertainty principle. It is called the rule of *commutivity*.

It deals with the order in which measurements are to be carried out, that is, the sequence of observations actually performed by the observer. Since each operator corresponds to an actual change in a physical characteristic, as, for example, the position or momentum of a particle, the question arises, "Will I get the same results when I observe first the position and then the momentum as if I observe first the momentum and then the position?"

We now know that you won't get the same results. The position and momentum of any object are complementary attributes. This means that knowledge of one renders the knowledge of the other as indeterminate, or unknown. This rendering, however, doesn't really erase the knowledge of the complement. It still exists but in an encrypted form.

How does one decrypt the information? In the case of position and momentum, the following takes place. If I observe, i.e., operate on any qwiff, with the momentum operator M as my first observable and next observe with the position operator P, I will produce an encrypted qwiff, one that appears random or incoherent. If, on the other hand, I were to repeat those very same operations in a reversed sequence (first P, then M), I will produce another encrypted qwiff, different from the previous one. If I next subtract the second encrypted qwiff from the first encrypted qwiff, the result will be, except for a constant multiplying factor, the original qwiff I started with. In other words, I have not gained any information by carrying out such a sequence of measurements and their reversed sequence. I end up with just what I started with, the original qwiff. This means that I still do not know the P or M of the object.

The particular sequence $MP - PM$ is the commutator of P and M. Whenever using the "commutator" produces the original qwiff, we say the observables P and M do not commute. This means that it is not possible to know both the P and the M of the object simultaneously. We can't gain this knowledge because a P observation encrypts M knowledge and vice versa.

A simple example of the noncommutivity of two operators can be found in the movements of a book placed upon a table. If you place a book before you, consider the following two rotations:

1. Turn the book clockwise one quarter-turn.
2. Turn the book up off the table toward yourself.

When you turn the book clockwise, imagine a clock face on the table so that you always turn the book, whatever position it happens to be in, from 12:00 to 3:00. Similarly, when you turn the book upward toward yourself, it can end up standing on its spine or side depending on how you start off the operations.

If you actually carry out this simple experiment, you will see that you do not end up with the book in the same position if you do 2 first and then 1 as you do if you use the reverse order.

What happens when two observables do commute? Then both attributes are capable of being known at the same time. For example, it is possible to know how far a particle is from a wall and from a ceiling enclosure. Or it is possible to know the momentum of a particle moving toward the wall and toward the ceiling at the same time. This latter case simply means that it is moving at an angle from one corner to the diagonally opposing corner.

If Pc and Pw stand for positions of the particle with respect to the ceiling and wall, and Mc and Mw stand for the momentum in the directions of the ceiling and wall, respectively, we will find that the P's commute among themselves and the M's also commute among themselves. This means, for example, the observation of Pc followed by Pw does not encrypt the qwiff. If I reverse the sequence of observations and perform Pw followed by Pc, no information is altered.

The commutator of Pc and Pw, $PcPw - PwPc$, in this case does not produce the original qwiff. It vanishes instead! A similar situation takes place if I commute Mw and Mc, Pw and Mc, Mw and Pc. In each of these cases an observation of one followed by the other gives exactly the same result as when these observations are carried out in the reverse order.[†]

The only way this can happen is that these observable-operators are compatible with the qwiff on which they operate. Or the qwiff can take on a form that is compatible with both observations. In other words, it is possible to know both observables simultaneously.

The qwiff has no objections to this added knowledge. We say that the qwiff "satisfies" both Pw and Pc at the same time.

Here another simple example comes to mind. Imagine that you are watching a game of jai alai (a game played with a tiny, hard ball and two players each wearing an elongated "basket" on one arm that is used to catch and throw the ball). Usually jai alai is played in a large rectangular room with one huge vertical wall as a backboard, the other walls removed so that spectators can watch the action (a net normally encloses the playing space to keep the ball from hitting a fan), and a rectangular floor approximately three to four times longer than it is wide. As you watch the match in your imagination, keep in mind the location of the ball with respect to the wall and the floor at the same time. Also keep in mind the momentum of the ball, how fast it is moving toward the wall (which you can determine by watching how far it bounces away from the wall after it is hurled against it) and how fast it is moving toward the floor (which, again, you can determine by how high the ball bounces).

Now stop the action by "freezing a frame" in the game. The ball is suspended in air. But where? To find it you must determine its Pw and Pf, its

---

†This says that if $PcPw - PwPc = 0$, then $PcPw = PwPc$.

distances from the wall and floor. Imagine stretching a tape measure from the floor to the ball and then from the wall to the ball. Now repeat, but in the reverse order. As you see, there is no difficulty with measuring one length before the other. It matters not which measurement comes first.

Next you want to determine by measurement the momentum of the ball with respect to the wall and floor. How can you do this? One way is to snap pictures of the ball and keep the lens shutter open long enough so that the ball makes a streak mark on the film similar to what you see on a developed photograph when you didn't hold the camera steady. By using two cameras, one facing away from the wall and the other facing from the floor, you will get two picture records, two streak images. The length of each streak tells you how fast the ball was moving parallel to the plane of the shutter. Again there is no problem in determining Mw and Mf, the momenta of the ball with respect to the wall and floor, respectively. The two observations commute; both are knowable at the same time.

It is even possible to measure both Pw and Mf simultaneously. The "streak-photo" recording of the Mf does not disrupt the efforts to record the exact position of the trajectory with respect to the wall. That's because no change in that momentum is introduced by the position measurement. A similar situation occurs with Pf and Mw.

Thus simultaneous knowledge of two observables, two possible attributes of a physical system, are knowable only if the operators representing the observables commute. Lack of commutivity means that the observation of one of the observables encrypts information about the other. When observable one is observed, its qwiff takes on a certain form. This form, like a coded message, contains within it other qwiffs encrypted. These other qwiffs are, like wave patterns, superposed or added together. It is their sum that gives the original qwiff its form. Conversely, when observable two is observed, the qwiff takes on the form of one of the previously encrypted qwiffs. In a similar manner this qwiff is made up or composed of a sum of qwiff patterns. Each of these now encrypted patterns is a possible pattern for observable one.

In brief, when observable one is observed, observable two is encrypted. When observable two is measured, observable one is encrypted. When two observables fail to commute, their commutator reproduces the original qwiff, i.e., nothing has been learned about the system by applying the operators to it. On the other hand, when two observables commute, application of these operators to the qwiff alters the qwiff so that something can be learned. The system changes to accommodate the observables.

Now, there is one additional complication to all this: time. In any real situation one observation is usually carried out after the other, meaning that a time interval exists between the observations. The question naturally arises of which observables commute through time? In other words, what can I know that won't change randomly and chaotically on me? What is securely knowable? This question is even more critical since time itself is not an observable.

There is one and only one observable that is generally able to withstand the tides of time, and that is energy. All other observables are subject to radical transformation if they do not commute with the energy operator. Because this energy operator is so special, it has been given a special name. It is called the Hamiltonian, after the nineteenth-century Irish mathematician Sir William Rowley Hamilton, who supplied a simple means to calculate the motions of complex classical mechanical systems by first determining their energies. Whenever a system could be described by simple mathematical expressions of energy, they turned out to be exactly solvable. Missile launchings are energy systems; so are clockworks and most complex machines.

Such energy systems were later called Hamiltonian systems. In them the complex equations of mechanics took on particularly simple forms, enabling the mathematician to obtain solutions to the equations of motion wherein certain variables of that motion were unchanging.

The key variable in Hamiltonian systems is always the total energy of the system under investigation. Often there are other variables, i.e., observables that also remain constant in time in these systems. When quantum mechanics was first developed, it was recognized that Hamiltonian systems were capable of being formulated directly and simply. It was the Hamiltonian that "ruled" the formulation. The big question in attempting to solve a problem in quantum mechanics is, "What is the Hamiltonian?" Once that question is resolved and the Hamiltonian for the system is recorded, all one needs to do is follow a simple mathematical procedure to obtain a solution for the motion. Now if other variables were constant in time in classical mechanical systems, it would usually mean that their corresponding quantum-mechanical observables would commute with the Hamiltonian. This would mean that these observables were also unchanging. They were simultaneously knowable with the total energy of the system.

Since energy is the key variable in determining the stability of a system through time, we can think of energy as a kind of cosmic, or universal, memory. In our minds lie the real energy that we see "out there." In energy lies our desire for stability, our need for the status quo.

Thus we come to how we shape and transform the physical world. We do this by our choices of observables. Hamiltonians are the operators we apply in our psyches in our everyday desires to maintain things unchanged. For example, the things in life we call "good," like "Mom's apple pie," are good because we remember "Mom." Our observables either commute with our memory records or they don't. These memory records form our Hamiltonian system. We remember "Mom" because something in us, which we call memory, is unchanged energy. In our recall we tend to recall what is pleasurable for us. We tend to use those variable-observables that allow energy to transform while remaining conserved so that "Mom" appears in different guises but she remains "Mom."

We are all part of the big "Hamiltonian." Each of us siphons off a piece of the "Hamiltonian flow." In our smaller universes we try to maintain constancy, but fail. We haven't taken into account all the variables. Maybe we want the world to be a certain way, conforming to desired expectations. We each have our "pet" observables. We each use our "pets" as filters on the outside world. By siphoning off less than we could we build up partial Hamiltonians, small images of the big world. We leave out essential interactions. With our smaller Hamiltonians we find that our "pet" observable-operators commute. By narrowing our possibilities for learning, our "pets" never change.

But our "pets" do not commute. It is an illusion that they do. Thus we are forced to reach out and touch, to reexamine our memories, our Hamiltonians. By including more interactions we find the world more complex. The old observables, those that we now see as failing to commute with the new Hamiltonian, we throw away or keep as approximate reminders of the past. By utilizing our new awareness, seeking new combinations of observable-operators, we find that we can change ourselves. In doing this we find that we change the world. This is evolution.

Bohr's Principle of Correspondence and the Masking of the Quantum Jumping Nature of the Universe

Classical physics really began with the early Greeks. Aristotle had already coined the word *physica* from the Greek word *physis,* which means spirit. In those days not much distinction was made between terms like *energy* and *force* Even in the time of Leonardo da Vinci (sixteenth century) it was thought that a falling body got heavier as it fell to the ground. I'm sure that for Leonardo, "heavier" meant it gathered more momentum, but the word *momentum* hadn't yet been invented.

As human ability to analyze grew and distinctions became more pronounced, the constancies of nature began both to appear and disappear. In Aristotle's time there was earth, air, fire, water, and the quintessence, ether — the spirit of life that pervaded everything. With Galileo and finally Newton these earlier constants of nature vanished, replaced by abstractions such as energy, momentum, mass, and force. And on top of all this was motion. How things moved and what caused them to move was for Aristotle no real problem, for there was, for him, a natural order to the five elements: Earth below water below air below fire below ether. Things moved when anything was out of its natural element. Thus water rose in the air when it was "mixed" with fire. The more fire the water held, the more that fire was out of its natural, elemental place. This pushed the water up until the fire left it, so that it would fall to earth once again.

In Newton's time, force was the cause that created motion. Once moving, an object continued to move. That was called having momentum. Changing momentum meant exerting a force. By applying a force to an object and causing that object to accelerate, work was done on it. With the recognition that more work meant more speed, the idea of conservation of energy became apparent. Work was recognized as a form of energy. If an object failed to increase its speed as a continual and constant force was applied to it, it would mean that another hidden force was acting and opposing the applied force. This was the undesirable force of friction.

For example, when a sky diver jumps from an airplane and assumes the flat-out "pancake" position in the air, he soon reaches "terminal velocity." The force of gravity pulling him down is countered by the force of air friction pushing him up. The result is that the diver falls with a constant speed. The work of gravity's pull is used on pushing air molecules around. Thus work was turned into heat energy, the movement of those air molecules.

The insight that energy was accountable for work led to the mechanical world view we have today. Machines are only useful if they aren't wasteful of energy. A badly designed machine means poor efficiency, which in turn means that too much energy is being drained by useless friction. Thus it was no wonder that energy systems, Hamiltonian systems, became theoretically important. With the discovery of the quantum the Hamiltonian became even more important.

In 1900 Max Planck created the theoretical concept that started the quantum revolution. He recognized a new form for energy. He saw that the equations that incorrectly predicted the color spectrum (i.e., the distribution of the light's intensity with color, so much for red, so much for yellow, etc.) could be corrected to conform to observation if he assumed that the light energy was emitted in units, or quanta. Each unit of energy depended on the light's color in a new way. Such a dependency of energy on color for light was unexpected. Indeed, there was no classical mechanical theory that predicted it. The formula expressing the relationship between the color and the light energy was, however, quite simple. Since whatever receives or is sensitive to light distinguishes colors through vibrational rates (our eyes tell the difference between blue and red because the electrons in our retinas vibrate at different rates), colors are measured by frequency, the number of vibrations per second. If f stands for frequency and E stands for energy, Planck's simple formula was $E = hf$.

The h in the formula was a factor never formulated before. Later we will say more about it.

Since red light had a lower frequency than blue light, this formula meant that red light had less energy in it per unit or quanta than blue light.

Later, Niels Bohr used Planck's simple formula to explain how an atom of hydrogen was able to radiate light. Bohr found that inside the atom a tiny electron was undergoing a sudden, discontinuous kind of movement, which he called a quantum jump whenever the atom emitted light. The difference in the electron's energy before and after the jump was found to be the E in Planck's formula, and therefore this energy was seen in the color of the light emitted.

73

This E was the first energy quantum.

With greater development, particularly from the post – World War I European physicists such as Louis de Broglie, Max Born, Niels Bohr, and especially Erwin Schroedinger, the importance of Planck's h factor became much enhanced. All physical processes depended on h. Nothing ever happened without at least one quantum of energy passing between interacting objects. Schroedinger developed an equation that described how Planck's h was used in all physical processes. Here, the Hamiltonian is found once again. However, instead of observables being simple numbers in an equation, they are operators, mathematical constructs whose only function is to change qwiffs in a prescribed mathematical manner.

Hamiltonians are energy operators that operate on qwiffs. Qwiffs respond to these operations and change, much as a vibrating string changes when a violinist touches the fingerboard of the violin.

In all of this development and much that followed it in later years, one principle, discovered as early as 1913 by Niels Bohr, remained true. It was called the *principle of correspondence*. We may think of it as the principle of "illusion."

To grasp what the correspondence principle says about the way we observe the world, I want to tell you about a motion picture that illustrates its many facets. It was called *The King of Hearts* and it concerned a soldier who finds himself in a small French village during World War II. Not speaking French well enough and fearful of appearing quite conspicuous in Nazi-occupied France, our English soldier seeks refuge in a French insane asylum. Upon hearing of the encroaching German military machine, the "normal" townspeople desert the town, leaving it in the hands of the "crazies," who manage to find the gate of the asylum open.

The inmates soon occupy the town and take it over. Now come the Germans, who are of course oblivious to all of this. When they arrive on the scene it all looks perfectly normal to them; after all, the French are certainly not as "normal" as the Germans who occupy them.

At this point we realize that the "crazies" are still crazy, but because they are now "townsfolk" they appear, to all intents and purposes, quite normal to the occupation forces. Even later, when the Germans are driven out of town by the incoming Allied forces, the "crazies" appear "normal." It is only when the real townspeople return to their village that the secret is out and the inmates are returned to the asylum. As I watched the film I thought that in many ways we, the people of earth, are the inmates. We are crazy and sane at the same time. We are allowed to play roles in life's great village, and as long as we continue our masquerades we remain "in character." It is our continuance, our wish for the continuity, that glues us to our destinies. Underneath it all we are all quite "nuts." No one alive can really take life seriously with its myriads of kaleidoscopic paradoxes; but we do.

Like us, the physical world is also a little "nuts." To bend the double entendre over, each "nut" is a quantum. Underneath the slick illusion of mechanical continuity and all of its causal consequences lies the great cosmic joker, God, and God's jokes are quantum-mechanical toys. And Niels Bohr figured out how God made the toys work. Bohr called his discovery the correspondence principle because he realized that our normal, or classical, world view is continuous. Yet his discovery of the quantum within the atom showed that atoms were fundamentally discontinuous in any transaction involving observers. How could the "atomic inmates" be so erratic while the "villages of atoms" that make up the macroscopic universe appear so normal and orderly? Bohr's discovery showed how the "quantum insane asylum" *corresponded* with the "normal atomic village," i.e., the orderly classical world of continuous motion.

To understand this we need to look at what physics means by the concept of "scale." A scale is a device used to measure something, like a bathroom scale or an architect's scale. When, for example, a new house is going to be built, a plan is drawn up. The plan is drawn to scale so that, for example, 1 inch represents 10 feet or perhaps even 200 feet. It is even possible that a scale model may be built before the actual house is constructed.

In many films utilizing special effects, scale models are used. The builders use model ships instead of real ones. For example, they place the model ships in a wave tank to simulate the ship's movements when it is at sea. The problem that arises is that if one shoots the scene in "real time," twenty-four frames per second, the scene would appear "hokey" and fake. No real ship would pitch and roll as quickly as a model ship does. The scale models, being lighter per unit of volume, move faster than a real ship. So the filmmakers must shoot in slow motion to make the scene appear real.

By scaling time in this way the filmmaker achieves realism, but it isn't perfect because of the waves themselves. They, too, have scale. A wave slosh is not the same for a tiny wave as it is for a six-footer.

Suppose we tried to build a scale model of an atom. This scale would have to correspond to perhaps two-billionths of an inch. Instead of scaling down as we do for a building, we must scale up. Expanding an atom to this scale is like imagining your thumb to be as large as the whole earth. How could an atom appear on such a scale?

The answer depends on what we mean by increasing the atomic scale. When Bohr discovered how tiny an atom was in reality, he realized that Planck's constant, h, played a major role in setting the atomic stage. Planck's h determined the scale of atomic phenomena. The atomic scale depended on a formula using h, the mass of an electron, m, and the electron's electrical charge, e. By combining these numbers Bohr calculated the atomic scale as h^2/me^2. Let's look at this formula for a moment. Each term plays a role in setting the atomic stage. The e measures the force of attraction between the atomic electron and the singly charged atomic nucleus. Since there are two charges present we have e^2. The m measures the mass of the electron slightly corrected to take into account also the much larger mass of the nucleus. Since the nuclear mass is at least 1846 times the mass of the electron, the electron is freer to move and thereby determine the range of the atomic stage by its wider roamings. Let's call h^2/me^2 by the symbol a_0 (a_0 is called one bohr by atomic physicists today).

If m were to somehow increase in magnitude, a_0 would actually get smaller. We can see this in the formula, and we can also understand this when we realize that m measures the inertia of the electron. Increasing m means making it more inert, less able to move about, and therefore decreases its range of space. Just as for a constant applied force, heavy things move more slowly and don't go as far as lighter objects, a heavier electron would also be so inclined. Especially if we were to increase e as well. By increasing the electrical charge the electron would be held by a stronger force to the nucleus. This, too, would restrict the electron's movements and thereby reduce the atomic scale. Thus both mass and charge are inhibiting factors in the formula for a_0.

On top of all this is h. By increasing it we expand the atomic scale. The h measures the uncertainty in our knowledge of the electron's location on the atomic stage. It is a tiny bit of chaos, a little craziness in our atomic "actor." It appears as h^2 because of the way energy of motion is represented in physics. Motion energy depends on the square of the momentum, which in turn depends on h.

So, to increase the atomic scale we can decrease e or m or increase h. In so doing we would of course be altering the delicate balance between these constants of nature. If h were just one thousand times bigger than it is, atoms would be one million times larger than they are. In such a world the chaos and quantum jumping would be overwhelming. For example, when an electron takes a quantum jump from its next to lowest energy state to its lowest energy state it jumps about three bohrs (three times a_0). For a normal h this is a distance of around six-billionths of an inch. If h were a thousand times larger, this jump would take place over six-thousandths of an inch. We can just begin to see this distance with our eyes. Objects would take on a very strange glow as a result and appear to be buzzy and fuzzy.

More than likely, the computer industry — especially with its emphasis on microchips and magnetic materials, which stack "bits" of information in spaces smaller than a thousandth of an inch — would only exist in our imaginations. A "bit" this size would be subject to random meanderings beyond the control of human beings. Even our thought processes would become jumbled. We would probably act quite "stoned" most of the time. This may be quite delightful for a short time but would after a while be quite disastrous, especially if you needed to learn and remember anything.

The h is the randomizing factor in the equation of life: Without h there are no errors and there are no surprises. The e and m are stabilizing factors of life: m makes matter inert, and e allows matter to interact electrically.

Without *e* and *m* every step on the surface of the earth would be a chancy affair with quicksand beneath our feet wherever we walked. Between *h* and the partners of inertia and electrical attraction, *e* and *m,* lies the whole universe and also all life.

So far, all attempts to "scale" an atom have failed. There simply is no way to really look at an atom. That is because what we use to look with always involves the three partners, *h, e,* and *m.* Any change made on one or more of the partners must be carried over to the observer as well. Thus a scale model of an atom cannot be built. But what about the atom itself? Do atoms ever increase their size? Can an atom be a scale model for itself? The answer is yes. Atoms do increase their sizes. They do this when they become excited.

The size of an atom is a measure of how far away from the nucleus an electron inside the atom can roam with any probability. When, for example, the electron is in its lowest possible state of energy, it is said to be in the *ground* state. The electron's qwiff is densely packed near the nucleus of the atom. The pattern appears as a cloud. Moving from the center out toward the "cloud's" edge, the pattern begins to disperse and becomes less dense. The range of greatest density is also the range of greatest probability for finding the electron. Even though the "cloud" has no actual final edge or boundary, it "more or less" is contained within the range of one bohr (two-billionths of an inch) or more.

A number can be assigned to designate this lowest energy state. It is $n = 1$. The *n* is called the atom's energy, or principal, quantum number. If the atom absorbs energy, it will necessarily change its principal quantum number. The *n* will increase to 2, 3, 4, . . . , on to infinity. A change of *n* is a quantum jump for the electron in the atom. If the electron jumps from $n = 1$ to $n = $ infinity, it jumps free of the atom. Any jump less than infinity keeps it bound to the nucleus, but in an excited energy state. Even though the jump to infinity seems enormous, it only requires approximately 13.59 electron-volts of energy to do it. This is the same amount of energy it takes to move an electron across a 13.59-volt battery or around nine D-size flashlight batteries (see aside at end of chapter). However, in making such a jump, the electron's qwiff becomes spread over a vast distance.

Even a modest jump from $n = 1$ to $n = 2$ absorbs about 10.2 electron-volts, three-fourths the energy needed to free it altogether. A careful calculation shows that the atom's electron "scales" according to n^2. Thus with $n = 2$ the atomic scale is four, meaning that the electron is likely to be found four times as far away from the nucleus as it was in the ground state. With $n = 3$, the scale increases to nine, etc.

A jump from $n = 2$ to $n = 3$ is also possible. The scale increases from four to nine, but the energy change is only 1.89 electron-volts. While a jump from $n = 1$ to $n = 3$ would take slightly more than 12 electron-volts of energy. When the electron finds itself in one of these higher energy, excited states, it has more total energy. It has taken on or absorbed the energy used to excite it. Surprisingly, though, it isn't moving any faster. Instead, the probability of finding it with a high speed or large momentum has actually decreased with the added energy.

The energy absorbed by the electron actually goes into increasing the electron's freedom. The nucleus holds it in and the electron's speed, or kinetic energy, tends to unleash it. Thus we find energy influencing the "dice throws" of the electron's life. By adding energy to a physical system such as an electron in an atom we are in a sense "fixing the horse race," influencing the outcome, and thereby changing the odds. In this example it is the odds of finding the electron at greater and greater distances from the atomic nucleus. From this peculiar quantum picture operating on the atomic stage, a new insight begins to appear.

The atom begins more and more to look like something real, i.e., appears to behave more and more smoothly and continuously, the farther away the electron is from the nucleus. Large quantum systems behave as if they obeyed classical mechanics. To see this more clearly let's compare two quantum jumps, a jump from $n = 1$ to $n = 2$ and a jump from $n = 100,000$ to $n = 100,001$.

The first jump absorbs more than 10 electron-volts of energy. The probable distance of the jump is around 3 bohrs, or about six-billionths of an inch. Although this is a tiny, tiny jump on the scale of inches, it is an enormous leap on the scale of bohrs. It is also a large change in energy.

The second jump takes place on a much larger scale. An electron with principal number 100,000 is likely to be as far from the nucleus as 20 inches! Imagine an atom of hydrogen floating in outer space and having absorbed just enough energy to make a transition from $n = 1$ to $n = 100,000$. That atom would occupy a space 40 inches across. Even at this vast distance the atomic electron must make a quantum jump if it is to absorb any more energy. When it jumps to $n = 100,001$, it changes its probable location by only .0004 inches, an insignificant distance when compared to 20 inches but a vast distance on the $n = 1$ atomic stage; for .0004 inches is 200,000 bohrs. The energy absorbed in this jump, enormous on the bohr scale, but insignificant on the inch scale, is .00000000000027 (twelve zeroes) electron-volts.[†]

And so we come to Bohr's discovery of God's illusion. A jump such as 100,000 – 100,001, although it is enormous for the electron, appears continuous and smooth on the scale of human concerns. Thus, just as a series of frames passing before our eyes at the rate of twenty-four per second makes a continuous moving picture, large quantum jumps look like continual motion.

Watching an atomic electron undergoing a series of transitions starting from $n = 100,000$ to $n = 100,002$, etc., until it reached ultimate freedom would appear as an object moving through space like a tiny satellite being

---

†Although such large atoms may appear to be mere fantasies, recent experimental research proves they exist!

blown by the wind. With each wind puff the satellite would rise, the wind energy moving it upward against the force of gravity. By adding more wind pressure we would get the impression that we were in control of this tiny voyager. Gradually, what is probable becomes what is actually happening, and with the appearance of reality comes the masking of the impossible.

With each atomic jump starting from the first, the size of the jump increases, the energy decreases, and the movement blurs into continuity. For the electron in an energy state with a principal quantum number of 100 thousand billion ($n = 100,000,000,000,000$), a jump to the very next energy state would take less than a mere wisp of energy to hurtle it through space a distance of over 6 miles. A 6-mile quantum jump seems like a lot for our little electron; but when n is that large, the atom is over 600 trillion miles in diameter. That distance is more than 100 light-years, or greater than the distance light can travel in one century. On the scale of our solar system, light has a much shorter trip. It only takes it eight minutes to travel from the sun to our earth, and only about five hours to continue the journey to Pluto. On a scale of 100 light-years, a 6-mile quantum jump is hardly worth noticing.

It is our interactions that determine our sense of normalcy. How we interact affects the scale on which we live our lives. It makes the scale in the same way that e, m, and h set the scale for our atoms. Altering one of these constants of nature would upset the delicate sense of our normal-appearing village. We are made of gigantic atoms 100 light-years in diameter. These huge atoms are cut down to size by our own presence. We interact. We are the atomic compromises. By narrowing our quantum jumps below the threshold of our senses we have a substantial and seemingly comprehensible world in which to live. As comprehensible as it appears, however, it is only an illusion.

As an aside, I thought it would be amusing for the reader to consider how tiny one electron-volt really is! A single 100-watt light bulb uses over 600 million trillion electron-volts of energy. However, it is only fair to tell you that quite a few electrons are involved in turning on that light.

In a 1-mil metal wire around 10 centimeters long there are approximately 200 billion trillion electrons! With a flip of the light switch each electron suffers the indignity of being pushed along by only three-thousandths of a volt of electrical potential. In an atom each electron undergoes a change of over 10 volts of electrical potential just to make the jump to the first excited energy state. Life in the metal-producing light at 100 watts for one second is easy living compared to making a jump in an atom.

On the other hand, in the example of the 100-light-year – diameter atom, the electron's jump took less than the thirty-millionth part of a trillionth part of a trillionth part of a trillionth part of an electron-volt of energy. Indeed a tiny whisper of energy to influence a 6-mile leap into the unknown.

PART TWO

$$\Psi * \Psi$$

NORMALITY IN
PHYSICS AND PSYCHOLOGY

What was considered to be normal and abnormal will change in the new psychology to come. Meanwhile the "new physics" sheds light on some old, perplexing concepts such as normality, mind control, order, disorder, observation, entropy, information, and the human condition.

3

How the Mind Creates Normality

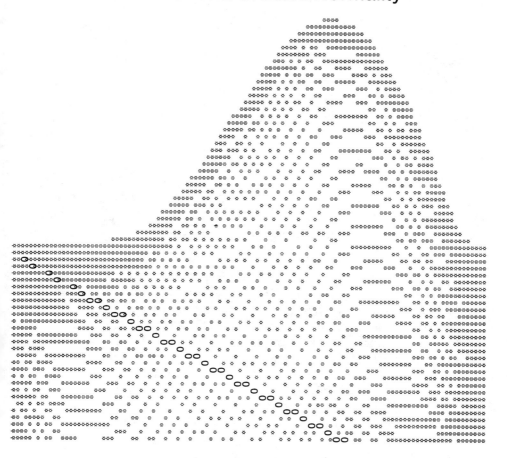

He was born to wonder about numbers.

*He balanced fives against tens and made
them sleep together and love each other.*

*He took sixes and sevens and set them
wrangling and fighting over raw bones.*

*He woke up twos and fours out of baby
sleep and touched them back to sleep.*

*He managed eights and nines, gave them
prophet beards, marched them into mists
and mountains.*

*He added all the numbers he knew, multiplied
them by new found numbers and called it a
prayer of Numbers.*

*For each of a million cipher silences he dug
up a mate number for a candle light in the dark.*

*He knew love numbers, luck numbers, how the
sea and stars are made and held by numbers.*

Carl Sandburg
"Number Man (for the ghost of Johann Sebastian Bach)"

"Normal" in a Sick Society?

Today we live under the powerful fear of nuclear destruction. This fear is virtually impossible to imagine in terms of its possible consequence, the total annihilation of all living beings. According to Dr. Timothy Hayes of the UCLA Medical School, this fear manifests itself in our everyday lives as a symptom that our minds can handle. In a recent talk before the Physicians for Social Responsibility in Los Angeles, he pointed out the documented case of a six-year-old child who on learning that his mother was dying of cancer refused to go to school the next day. It seems the child had developed a rather severe phobia: He was afraid to go to school because of the dead birds he might find along the route he took to his classroom.

Hayes pointed out that the child had substituted for the unmanageable death of his mother the manageable and totally rational experience of the discovery of a dead bird. The kid could handle dead birds. The death of mother, however, was unthinkable for the child.

Nuclear annihilation is unthinkable for most, and perhaps all, of us. Yet the fear is real. What "dead bird" can or do we use as a substitute? Perhaps our overemphasis on health? Our abusive attitude toward drugs and drug offenders? Our rationalization of meaningless jobs? And so on. Perhaps it is not an overstatement to say that each of us feels uncomfortable today for some rather apparent reason. This discomfort or indeed this unnecessary suffering, which often manifests as fear or even paranoia, may be traced to the tiny nucleus of the atom that lives inside each of us. Our fear of nuclear death is part of us. It lives in each nucleus.

Even our economics must be affected by such concerns. What we presently spend through taxes to support the arms race, particularly the nuclear arms race, must eventually lead to our own destruction. It is a paradox that our minds cannot or will not handle. Thus we may feel ill, have headaches, backaches, cancer, watch inane television shows, have bad jobs, flat feet, not enough money, bad food, too little exercise, too much exercise, and bad breath because we can handle these problems. The real cause of our discomfort remains hidden by its paradoxical nature.

Our rationalization or resolution of the paradox is that we must be strong to frighten our enemy away. Surely no rational enemy will attack us because we have a bigger stick than he has. But, using quantum thinking for a moment, let's posit ourselves as our enemies. If you had a big stick-wielding bully breathing down your neck, what would you do? Your fear is, of course, that he will hit you. After all, why else would he carry such a big stick? You would — I'm certain that I would — plan to build a stick at least as big as the bully's, maybe even bigger.

Further, there is a subtle trap to the story. He who carries a big stick is a bully. He certainly is neither a fun-loving nor a "love-loving" individual. You cannot love and fear at the same time. According to the quantum model of consciousness, love and fear are orthogonal to each other. To possess one is to lose any value for the other.

If we do live under such threatening considerations as nuclear disintegration, then what do we mean by the concept "normal"? What does it mean to be a normal person today? I've asked people what they think it means to be normal today. The responses are quite varied. If we are to move toward a new, more effective psychology — indeed we are doing so at this very minute — it will be necessary to determine "once and for all" the meaning of normality. To do this we need to look toward mathematics and how the mind creates "paths of least action" or, as Rupert Sheldrake calls them, "chreodes."

First, according to the mathematics known as statistics, *normal* refers to a particular point on a particular kind of curve. It is this "curve" that we will attempt to follow in the next section.

Normality in the Center of a Bell-Shaped Curve

In mathematics a curve defines a precise relationship between two quantities called variables. A curve says that one variable "depends" on the other. We call the variable that is dependent the dependent variable, of course! The other one is called the independent variable. Usually we graphically depict this dependence by a drawing in which the vertical scale distance marks the value of the dependent variable and the horizontal scale marks the independent variable. Often the vertical variable is labeled y and the horizontal variable x.

We then say that we have a mathematical relationship, or "functional" relation, between y and x. The y is said to be a function of x, or we say that y depends on x. This means that if a value for x is given, we can see that a value for y is also specified correspondingly. We see this as the curve of y versus x.

Before we look at why the curve appears as it does — that is, how it takes the shape of a bell — I want to point out certain of its outstanding characteristics. The exact center of the curve, the point that marks the peak, is called the normal. Thus if we take as an example the number of people who have a certain body temperature versus the temperature scale, we have a typical bell-shaped predictive statistical situation.

Obviously everyone has a body temperature, even a dead body. Suppose we go to the city of Los Angeles and walk into a typical medical clinic. As you are well aware, you can't even talk to a doctor there unless your temperature has been taken. I'm sure you know that "normal" temperature is 98.6 degrees Fahrenheit. But if you ever actually measure your temperature when you are not sick you probably find that it is rarely exactly 98.6 degrees.

Suppose we take a thousand temperatures at UCLA's medical clinic one rainy day in L.A. Next let us "break down" the scale of temperature into temperature quantums. Each quantum will be one-tenth of a degree wide. We can think of a person's temperature as being made of so many quantums. For example, suppose my temperature was 98.45 degrees and yours was 98.48 degrees. I would have 980 plus 4.5 quantums. You would have 980 plus 4.8 quantums. But there are no fractions of quantums. So we would be considered as actually having the same temperature, namely, 984 quantums. As long as a temperature was greater than or equal to 98.4 and less than 98.5 it would, according to our rule of one-tenth degree equals one quantum, be considered as 984 quantums of temperature.

Now we look at all the recorded temperature data for these "typical" people on a rainy day in L.A. Starting with 950 quantums and continuing to 1100 quantums probably covers the range of typical temperatures. As we expect, the normal temperature is roughly 985 or 986 quantums. We see that what that means is there simply are more people with a temperature in the range 98.5 degrees to 98.7 degrees than in any other quantum region, say, 98.3 degrees to 98.5 degrees.

As we move away from the normal or, as statisticians say, "deviate from the norm," fewer and fewer people are found. Virtually no one has a temperature of 110 degrees and no one has one of 95 degrees. But how far can a person's temperature deviate and yet still be considered normal? Actually the answer is zero. However, we know that such a rigid limit is unreasonable. There are nearly as many people with 98.4 as there are with 98.6.

In other words deviations from the normal are normal! If we count the number of people who have temperatures outside the range of 96.8 degrees to 100.5 degrees, we would find a significant number. (I would guess nearly 333 people out of 1000 would fall outside this range.) But now suppose we sample temperatures from 10,000 people.

It turns out that from such a large sample fewer and fewer people per capita will be found outside this range. In other words in a sample of 10,000 people there will be nothing at all like 3333 people with deviant temperatures! Instead, around 1000 people are likely to be found outside this range. With an even larger sample of 100,000 people, about 3333 people will be found with temperatures outside the range of 96.5 degrees to to 100.5 degrees, or only 1 person in 30. (Compare with the sample of 1000 persons. There, 1 person in 3 was a "deviant.")

This rather dramatic squeeze toward the middle is not due to anything "psychic" or contagious going on between the people. Instead, it has to do with a peculiarity of the laws of statistics. Known as the law of large numbers, it says that as more and more cases are included in the sample, less and less likely will a deviation from the normal occur.

This squeeze toward the middle is shown quite clearly in the coin flip. What can a gambler expect for a run of heads if he flips his coin a certain number of times? In other words, how often will heads occur as he increases the number of tosses?

In four tosses he can expect to see H H H H, H T H H, H T T H, H T T T, and T T T T. There is only one combination of four heads, while there are four ways to have three heads and one tail: T H H H, H T H H, H H T H, and H H H T.

For two heads and two tails there are six combinations, and for one head and three tails there are four combinations. It is no coincidence that there are more ways to achieve two heads and two tails than any other distribution and that the average number of heads in four tosses is two. Yet in an actual game of four tosses we wouldn't at all be surprised to see a run of three or four heads occur.

With 16 tosses we rarely see fewer than 6 heads or more than 10. With 64 tosses any number from 28 to 36 heads appears normal, with anything outside this range rarely occurring. With 100 tosses the range is 45 to 55 heads; with 400 tosses the range is 190 to 210 heads; and with 900 tosses the range is from 435 to 465 heads. In the latter case, the number of times a run contains more than 480 or less than 420 heads is virtually zero. With each increase in the number of tosses, the number of deviations from the average number, which is one-half the number of tosses, becomes a smaller and smaller proportion of the number of tosses. With each increase in the number of tosses, there are more and more ways to achieve the normal condition. With 900 tosses of a coin there are over 10^{272} (10 multiplied by itself 272 times) combinations of exactly 450 heads, while a run of 900 heads has only one combination.

If we make a plot or graph with the number of times a run of 450 heads occurs in a toss of 900 coins (note the run of heads does not necessarily occur consecutively; indeed such a string of heads would only constitute one possible way for 450 heads to occur) and then add to the graph plots with 449, 448, . . . ,2,1, as well as runs of 451, 452, . . . , 899, 900, we will obtain the famous bell-shaped curve of statistics shown in Figure 22.

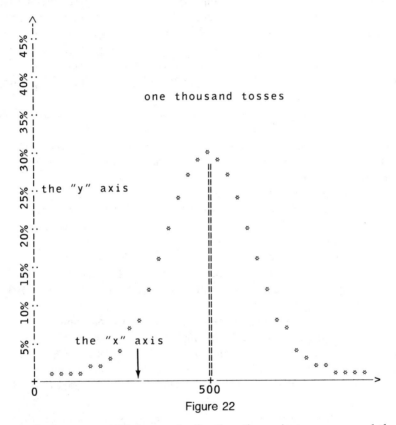

Figure 22

Similarly, we could draw graphs for the other coin-toss cases, and these are shown in Figure 23. With only four tosses the curve is not "curvy" but appears as a series of steps or quantum jumps. As the number of tosses increases, the jumps appear smoother because the numbers are getting larger and the scale of the graph is getting smaller. As we reach the domain of large numbers, smoothness and predictability overcome roughness and discontinuity. This is quite similar to Bohr's correspondence principle discussed in the previous chapter.

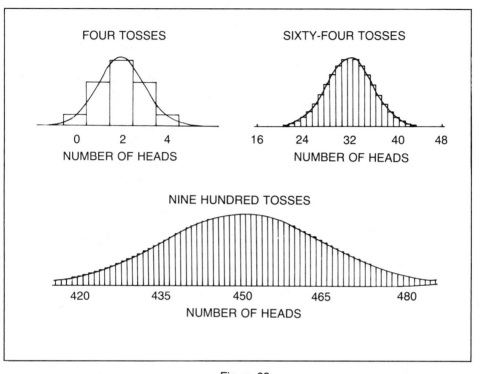

Figure 23

The lesson for us is that anything repeated enough times will eventually become boring. It will lose the surprise of discontinuity. Gradually, with ever-increasing numbers any result will deviate less and less from the average. The average will emerge triumphant — the meek shall inherit the earth. Gradually the law of large numbers, the squeeze toward the middle, creates the disappearance of individuality; we become blank faces in a crowd.

The loss of the quantumlike discontinuous rough edges is replaced with the "ring" of the bell-shaped curve of predictable, and therefore respectable, behavior. The crowd obeys the inevitable law of large numbers. This inevitability holds for all countable events. It matters not whether we are counting bodies of humans with varying temperatures or if we are looking at the firings of neurons in our own nervous systems. A normal behavior will emerge if enough events are counted or observed. Although each event may be remarkable, the average event is dull, repeatable, and secure.

The Bell-Shaped Curve and the Human Mind

Statistics influence us because we are composed of large numbers. There is hardly anything we do that doesn't involve a large number of events occurring simultaneously or within a small interval of perhaps only a few thousandths of a second. There are over 10^{12} (10 multiplied by itself twelve times) neurons in our human nervous system. Each of these neurons activate around ten thousand feedback loops in connection with other neurons and the outside environment, muscles, etc. Each thought may be composed of one billion neurons firing, with each feeding back ten thousand messages to the environment and to our eyes, fingers, legs, tongues, and other sizable organs.

In a recent book, A. Harry Klopf, from the Avionics Laboratory of the Air Force Wright Aeronautical Laboratories at Wright-Patterson Air Force Base, Ohio, discusses the idea that we are made of hedonistic neurons and that these are responsible for our memory, learning, and intelligence. Klopf differentiates two conditions of the human nervous system: homeostasis and heterostasis. Homeostasis refers to a condition where the system attempts to reinforce or maintain a steady or "average" behavior. Such behavior may be quite consistent with the survival of the person, for example. Heterostasis is defined as the seeking of a goal or maximum condition, a standing above the crowd, so to speak. It is our homeostats that make us sit at home watching TV while our heterostats turn us into television performers.

Homeostasis operates through the laws of large numbers. Heterostasis doesn't; it abhors the masses. Thus our very thoughts are influenced by this dichotomy. Our expectations are usually an outcome of the averages. Our hopes and dreams are quantum events of heterostatic activity.

We will return to Klopf's ideas in Chapter 7. It is enough to point out here that with such a large number of mental events occurring each time you raise an eyebrow in incredulity, it must be a bell-shaped game. Not all of your neurons, muscle fibers, skin patches, and nerve endings want to go along with your incredulous sneer. Some, undoubtedly, want to laugh or even inhibit the actions of the other components composing the sneer. But the majority wins because it not only outweighs the minority but does so in many more different ways.

In a society of sneerers, your sneer is expected. You have learned well how to sneer. You have watched your peers sneer. You have learned just how to hold your head, to flare your nostrils, and to condescend. The society of sneerers could conceivably encompass a whole country, perhaps France or England!

Sneering becomes the accepted, expected norm. Your face is the face of a nation.

Speech patterns, too, are reinforced with statistics. Japanese people are notorious for the inability to pronounce the letter *r*. Their *r* is more like our *d* but not quite. It is quite distinct from their letter *d*. However, Japanese children raised in an English-speaking country have no difficulty with *r*. Similar patterns of speech can be found in other language comparisons. People are the same but their environments change their acceptable "mouth noises."

The "great desire" of normality — the squeeze towards the middle — is reinforced as each person learns the acquired characteristic. As it is with nations, so it goes with neurons. It is not that a sneering nation is compelled to sneer by genetic engineers who have mechanically contrived to produce the sneer. Instead we have an alteration of probabilities for each neuron so that a neuron is more likely to produce a muscular contortion resulting in a sneer than in a not-sneer (i.e., a smile).

With each alteration in the probabilities there is a shift in the average. We can see this in Figure 24, which contains plots of probability in terms of percent per standard unit versus the number of events and/or number of standard units.

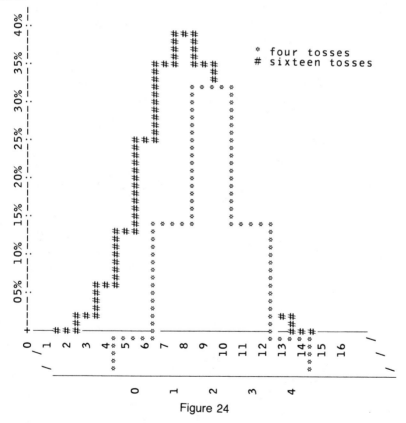

Figure 24

By altering the probabilities we are able to shift the distribution function. In the above figure we are looking at a loaded coin flip in addition to the usual fifty – fifty dichotomy where the probability of flipping a head is 0.5, or 50 percent. Here we set this "loaded" probability to 7/8, or 87.5 percent. The average or expected result is now fourteen out of sixteen instead of eight out of sixteen.

In a similar manner an acquired characteristic is an alteration of the probability patterns of neuronal firings. It is here that the quantum wave function "does its thing." By altering the boundary conditions of its existence (see Chapter 1) we are changing the probabilities of neuronal firing patterns.

Thus it is that the "mind" alters qwiffs, which as a result alters the probabilities of nerve firings, which in turn alters the behavior of the living system. Habits are slow to learn and slow to lose because of the law of large numbers.

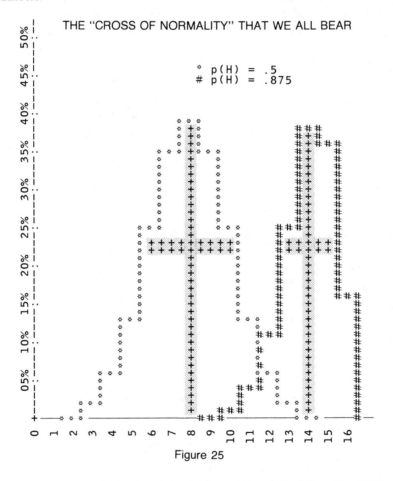

Figure 25

In Figure 25, which I label the "Cross of Normality That We All Bear," we see how the statistics change as a characteristic is acquired. Although we are only looking at sixteen events, this is enough to make the point. The first "curve" (of *'s) is centered or normalized to eight events. This is the situation in, for example, sixteen coin tosses with a fair coin. The vertical bar marks the normal, and the curve, which consists of a series of symmetric steps centered about 8, gives the probability in the number of "successful" events at the foot of each step. Thus, for example, to have five successes, the probability is roughly 13.33 percent, six successes gives 24 percent, etc.

The horizontal crossbar marks a length of two standard deviations, which in this case is from six to ten and therefore four "successes" wide. Within this crossbar, 67.5 percent of the results of sixteen coin-toss games will lie. In other words, the chance that you will have from six to ten heads in sixteen tosses is about 67.5 percent. Moving outside the "cross of normality" only occurs 32.5 percent of the time, and most of that is "gobbled up" by only one standard deviation on either side of the cross. Thus the chance you will have from four to six heads is about 16.25 percent, and about 16.25 percent that you will see ten to twelve heads. Three or thirteen heads occurs less than 2 percent of the time. Fourteen heads virtually never occurs, with a probability of only 0.3 percent.

Suppose that we are looking at sixteen neural events that compose the tiniest flicker of a sneer on the edge of your nostril. With such a fifty – fifty probability distribution, no one could tell the difference between your sneers and your smiles. Suppose it takes at least thirteen "successes" to make a sneer. Eight successes just puts your nose under stress! If you couldn't alter the probability of a single neuron "creating" the sneer profile, you would be doomed to sneerlessness.

Indeed, in such a situation your sneerlessness is your homeostasis. Rarely, perhaps in 1 or 2 out of 100 trials, you would successfully show the nostril flair of a true sneer. Perhaps you were watching yourself in a mirror when this rare occurrence took place. You were conscious of your sneer. You even cheered for your sneer. At that moment consciousness entered the game and altered the crap tables of eternity. A small event, no less significant than the grandest, the probability of a single happening, a single neuronal firing, was instantly "collapsed." Instead of that event only occurring with a probability of 50 percent, it now occurs, for example, with an 87.5 percent probability, or seven times out of eight.

Your awareness was the needed feedback for heterostasis to occur. You have learned something. Indeed, your neuron has been trained to sneer. Now the game is suddenly shifted. There is still a cross of normality, only this time the normal behavior is marked at fourteen event-successes. The horizontal bar still stretches across two standard deviations, but this time it is shorter. It is only 2.64 "successes" wide. With success, little tolerance is permitted. Now any number from thirteen to fifteen successes results in a clearly marked sneer. And this occurs about 67.5 percent of the time.

There has been a squeeze toward the middle. With more and more events taking place, there is less and less tolerance for error.

Thus with an increase in the probability of the event, clearly we would expect more such events to occur. That explains why fourteen is now normal and eight is unheard of. But why is the tolerance of error cut down so much? If the probability for a single success was increased to one, so that every toss produced a success, then it would be impossible to have a failure. There would be no room for it, hence no tolerance for error. The curve would appear as a single vertical bar erected at sixteen events. Similarly if the probability for success was cut down to zero, we would find a vertical bar at 0 events. As the probability for a single success increases from zero to unity, that vertical bar sweeps across the graph from 0 to 16 and the horizontal crossbar grows in length from no events to a full four events at eight successes. From 8 to 16, the bar diminishes to zero length once again. It seems that nature and statistics give a wide berth to pure chance but tolerate little error once something is found to be a success or a failure.

In this way our neural nets all start off in the "tabula rasa" or clear-slate condition. In the middle of it all, with success a fifty – fifty crap game, the tolerance for improvement or failure is greatest. But the price we pay for succeeding or failing is that there is now an expectation put upon us. *Failure breeds failure, while success breeds success.* Although not guaranteed, failure comes harder once you have succeeded than it is to succeed in the first place!

And, by the way, "hard" simply means the probability is small.

How does observation change probability? What observes what? To answer the first question I want you to do a tiny demonstration. You will need another person for this.

Now that two of you are present, take a coin and flip it so that it spins in the air. Then catch the coin so that even you do not know which side of the coin is facing you. Now ask your partner, "What would I say the probability is that the coin now hidden under my hand has heads facing up?" Wait a while for your partner to catch on to this tricky question. Remember, you are not asking about the coin; you are asking about your knowledge about the coin.

In a while, your partner will answer "fifty percent." I'm sure that you agree with this utterance for, since you do not know which side of the coin faces you, you have no idea what to expect. Since the coin is a fair coin, both heads and tails are equally likely. Now, however, while your partner is watching you, peek at the coin, observing which side actually faces you. Repeat the question: "What would I say the probability is that the coin now hidden under my hand has heads facing up?"

Of course, the situation has changed, for now you know which side faces up. The answer can no longer be "50 percent." Your partner will probably chuckle knowingly and say something like "one hundred percent," but this isn't right either. Or he or she might say, "It's either one hundred percent or nothing," which is quite correct.

Your knowledge of the situation has changed the probability instantly. By becoming aware, you have altered the probability.

I'm sure some of you realize that this was a tricky kind of situation. Only your "mind" was changed by the knowledge, not the actual coin itself. You were only observing the coin, so changes in you shouldn't count.

But let's carry this one step further. How do you know that the coin is really heads up (if it actually was when you saw it)? You knew because you knew what heads looked like. If you couldn't tell the difference between heads and tails you would need to learn that difference. To acquire this knowledge your neurons must be involved in a learning game of pattern recognition. By repeated trials, the pattern is reinforced.

It is only through your ability to recognize the pattern that the reinforcement occurs. Failure to do so results in error. To learn the pattern of the difference between heads and tails, you must alter the probabilities of your neuronal firings. Consciousness enters into this. Its function is simply probability alteration through pattern recognition.

These alterations of neuronal firing probabilities produce the experience of normality. Normality may indeed be in error, but our thoughts follow the paths of normality. These paths begin with objects that are remote from our nervous system, relatively uncoupled except through remote interactions. These remote interactions constitute the outside world. In the next section we will look at how our internal "computer" works.

Artificial Intelligence Versus the New Physics of Consciousness

The latest vogue in modern psychology is to use the language of computer technology to describe the human mind. According to Jonathan Miller and others he interviewed, psychologists have come to the position that somehow our minds have memories that contain representations of the outside world. This is a far cry from the behaviorist schools, which see the mind as virtually nonexistent and all thought and behavior as conditioned stimulus-responses to environmental changes.

The vogue is catching on. Even computer technologists use psychological modeling to represent the memory of a computer. Neal Coultier describes the pioneering work of M. H. Halstead on "Software Science," which is an "experimental and theoretical discipline concerned with measurable properties of written material as expressed either in computer programs or prose."

Thus cognitive psychology, human-memory models and human-memory searches, long-term memory and short-term memory, and programming time are part and parcel of computer technology today.

Artificial-intelligence "freaks" abound. Their heartfelt dream is to make the first intelligent "humanized" computer. These specialists believe that all thinking, dreaming, and altered states of consciousness are programmable and mechanically creatable. But are they correct? In this section we will look at this possibility.

95

One of the primary concepts in the new physics of consciousness is that of "feedback." Through feedback, two processes occur:

- Self-reference
- Renormalization

Through these two processes probabilities are altered and a sense of "self" separate from the ONE arises.

The basic feedback is, however, bizarre. It occurs as a loop in time. Without this loop there would be no "I" sense and no basis for pattern recognition. This loop begins with an event, as clearly defined as possible in quantum physics, NOW. The qwiff then flows to a future event, one of several possible future events that have been set up by the NOW event, and then flows backward through or against the time stream, creating a disruptive disturbance. It is this interaction between the two streams,

$$N \rightarrow F \text{ and } N \leftarrow F,$$

where N is the NOW event and F is the future event, that "pops" the qwiff as described earlier in the book. The result is a sense of "I" and an intuitive apprehension or anticipation of F.

Thus is born the personal "I" sense, the experience of fear or anxiety *(angst),* the sense of continuance, and the realization of the "I" event as a pattern with respect to the future event F.

Now it is commonly thought that the sense of presence or identity we all feel arises from our memories; that we scan our memories for past events that help us define what to do next. Indeed we do. But it must be remembered that the "memory" we scan exists NOW. We aren't literally going back in time for answers to our present problems. (We can send our qwiffs back in time by remembering to pop past qwiffs, however.) I am claiming that we actually look toward the future through this qwiff $N \rightarrow F \rightarrow N$ process, acting as if F is our memory in the future.

When this double-back flow occurs, a path between N and F is created so that F then also occurs. Every event NOW is connected to a past event, which the NOW event in a certain sense caused, and a future event, which from the NOW perspective is only probable but is certain from the F perspective. In other words, when an event becomes certain, it is NOW and vice versa.

All events that remain undefined through this loop are the great collective unconscious — the mind of God if you wish. These connections define our purpose and our "ourness." Thus suppose there are events $F1, F2, \ldots, Fn$ possible. From the N event perspective, N attempts to connect with all of these future events. Some of the F events are particularly similar; they differ only in small details and appear overwhelming. Without any discernment between the details of the future events, without any attempt to clarify where, what, when, etc., these events occur, the feedback to the "I" event is from all of the F events. This results in the sense of destiny.

One of the possible F events will occur. It will become the new N event. The next set of F events will appear. But this time the F events will differ because they are a realization of the new "I" event, the "I" – F possibilities. A refinement has occurred. With the realization of another F event, a further refinement will occur.

To see this in a model example, I have chosen a simple problem from mathematics. The problem is for x to obtain an identity. All x has to go on is that it exists in the future as the number 1 divided by itself plus 1. Or,

$$x \;=\; 1/(x+1)$$
$$\text{past} \;=\; \text{future}^\dagger$$

In this model both the processes of self-reference and renormalization are going on. In attempting to define itself, x (the unknown appearing on the left-hand side of the equal sign) "looks at itself" in a kind of mathematical mirror. It sees itself on the right-hand side of the equal sign in a new form as 1 divided by itself plus 1. By just guessing a value for x, it tries to fit into both expressions.

As it continues by always reaffirming itself in the future, as $1/(1 + x)$, it begins to converge on a value and renormalizes itself.

This is a self-referential equation. It says that x, which is a symbol for any number, is the same as the number 1 divided by the sum of 1 and x. Now consider the right-hand side of the equation as the future and the left-hand side as the present, or NOW. The NOW is attempting to define itself by the future. But once this equality is demanded, a restriction is imposed on the variable, x. The x must now *be* something.

A dynamic process starts off as x tries to fulfill its potential by becoming that which has been imposed from the future. The process goes something like this:

What am I? Am I 50? Am I 35? Maybe I'm 100? Let me try x is 50 and see: Does

$$50 = 1/(1 + 50)?$$

No, of course not. What do I do next? The equation says that on the right-hand side of the equality there is 1/51, a small number that is equal to .0196. On the left is my starting NOW number, 50, a number I guessed at. It didn't work.

But why didn't it work? Let me look at the right-hand side, only this time with x equal to .0196. Does this work? Does

$$.0196 = 1/(1 + .0196)?$$

No again. The right side, 1/1.0196, is equal to .9808. This is bigger than .0196 but smaller than 50. Perhaps the range of possibility is getting smaller. Perhaps

†Note to computerists: This is not a computer-language equation that says replace x by
    1/(1 + x).

ENTER THE VARIABLE

1 is approaching a real value. Let me try x as equal to .9806 as the future. Does

$$.9806 = 1/(1 + .9806)?$$

No again. The right side, 1/1.9806, is equal to .5049. This is smaller than .9806 and the range of possible values for x is smaller than before. Let me try x equal to .5049 on the right-hand side of the equation. Does

$$.5049 = 1/(1 + .5049)?$$

No again. The right side, 1/1.5049, equals .6645. This is bigger than .5049. Let me try it. The x is .6645. Does

$$.6645 = 1/(1 + .6645)?$$

No again. The right side, 1/1.6645, equals .6008. This is smaller than .6645. Let me try it. The x is .6008. Does

$$.6008 = 1/(1 + .6008)?$$

No again. The right side, 1/1.6008, is .6247. This is bigger than .6008. Let me try it. The x is .6247. Does

$$.6247 = 1/(1 + .6247)?$$

No again. The right side, 1/1.6247, is .6155. This is smaller than .6247. Let me try it. The x is .6155. Does

$$.6155 = 1/(1 + .6155)?$$

No again. The right side, 1/1.6155, is .6190. This is bigger than .6155. Let me try it. The x is .6190. Does

$$.6190 = 1/(1 + .6190)?$$

If you look at these attempts of x to gain identity, you will see that while at first it seemed hopeless, there does seem to be a progression taking place. By x trying to identify with a distorted form of itself, $1/(1 + x)$, x is rapidly converging on a number. It didn't matter what x was at the start. It has been renormalized by this process. If we were to continue this forever, we would eventually find that x could approach but never reach an exact value. At some point the process would stop. We would say that this point was "good enough for all practical purposes." Therefore, x is .61803 for all practical purposes.

Thus self-reference is like this. It is an ongoing process of reappraisal. It stops when it is "good enough." The same is true with our attempts to identify, to form egos. Each step is an attempt to be that which we cannot be, some form on the outside world or some form within ourselves, an ideal image, a hero, John Wayne, Marilyn Monroe, "Guru Joe," or "jolting" Joe DiMaggio. The point is that the identity is not in x but in the process by which x attempts to project itself into that which it isn't.

By this process, the future and the present "handshake" across time. Each trial value of x led to a further refinement, with a convergence or renormalization-identification-realization in consciousness. Since this is a process, it is dynamic. Since it is dynamic back and forth across time, it cannot be "seen" as a physical process. Physical processes occur only along one time direction, from the present to the future.

All artificial-intelligence devices are based on memory — only storing programs and data bases loaded from the past. The computer has no sense of the future. It uses the past programs as a basis for what to do in the future. It can learn to modify its behavior only according to a program that tells it how to accomplish this. But this program, too, comes from the past.

In the quantum-physical model of consciousness it is the future that decides the present! The past falls under control of the present! Shades of Orwell's *1984*. (Perhaps another synchronicity occurs as I write these words in 1983 for publication in 1984.) From the present's point of view, the future is only imaginable as a probability wave. The past is re-membered, re-put together, re-assembled, re-built, re-created as a real, past, absolutely fixed-in-the-mind event. Computers, artificial-intelligence devices, and certain robot-ized individuals known by their stick-to-the-rules philosophies have built-in programs to tell them what to do in novel situations. At least they do some-thing. These "individuals" are intelligent only so far as the past forms the only basis for the individual's present actions.

"Human BEINGS" are guided by a sense of their own evolved identities in the future. That is why humans don't seem mechanical. And they're not, in any "normal" sense of the meaning of that word.

In the next section we will look at why it is natural to be unnormal and the origin of creativity.

Why Unnormal Is Natural and the Nature of Creativity

Computers follow orders. They can't act at random and function in any way. There is really no bell-shaped curve of artificially intelligent behavior. A random computer is no computer at all. Random-number generators exist in computers. However, they are simply following the orders of an algorithm, a program that produces a string of numbers in a predetermined manner that "looks" random but isn't.

If you remember the example I discussed earlier in this chapter and look again at the "Cross of Normality" (Figure 25), you can see that as the probabil-ity of an event becomes certain, the curve collapses to a single line "100 percent long" without any width. Normal becomes perfect, with no room for error or creativity. Here the dividing line between human and mechanical behavior exists: All machines strive for perfection. To the extent that they work well, they must minimize all error.

As a child I stammered. My ability to speak certain words, particularly consonants, was impaired, perhaps for psychological reasons. To overcome this difficulty I learned to creatively substitute other words for the words I couldn't pronounce. In a sense my speaking improvement was a broadening of the bell-shaped curve; the more word substitutions, the broader the curve.

Creativity demands error. Through error, choice exists. Although all of us desire perfection in our lives, such perfection, if reached, would overwhelm us. Life, to say the least, would be horribly boring. Worse than trying some-thing and making errors is not doing anything for fear of failure.

Nature, through the action of the tiny quantum, is prodigious in making errors. It does not work toward perfection; she strives to create all things possible. And even if those things are impossible, she finds a way to do them too! Left to her own makings, nature will seek all possible paths to chaos. Once restrictions or boundary conditions are imposed, some paths become more likely than others. In other words, the mind of nature — or, if you wish, God — alters the probability of events occurring by the simple acts we call observa-tion. We are, life is, the observer in the game of the physical world. Without life, there would be only chaos. Life, through observation, imposes the limits of possibility and perfectibility. Life is the only crapshooter in God's cosmic dice game. Life and mind are synonymous. The mechanical order we seek and find in both living and nonliving matter is in our minds.

It is a remarkable paradox that life, in seeking perfection, something that nature would never attempt to do, actually seeks death. I mean this in the sense that mechanical perfection, i.e., movement without choice, without room for error, is a kind of death.

In the now classic confrontation between man and machine, such as exemplified by the television program "Battlestar Galactica," we have Adama leading his people across space while escaping from the dreaded Cylons, intelligent robots who have been programmed (by Satan, it is strongly hinted) to seek out and destroy all human life forms. Of course, the humans triumph in each episode. It is not their superior intellect that wins out. It is more the fact that the Cylons move mechanically, unable to anticipate the random movements of the humans. The Cylons are perfect but no match for the imperfect humans, who win out through their unanticipated actions. The Cylons don't care whether they live or die. They are incapable of imagining their own deaths — of course, because they are already dead.

In an unusual recent book, *A New Science of Life,* Rupert Sheldrake describes the nature of morphogenetic fields. These fields, Sheldrake believes, are responsible for the remarkable repetition of life forms throughout time. They exist as probability patterns extending in space much like the magnetic field surrounding a current-carrying wire. These fields "order" and shape matter so that matter will follow the pattern of the field. These fields are capable of changing in time as nature "learns" how to perfect the patterns. In this way the tiny newt grows a new eye and the lizard a new tail after these organs have been lost to predators, for example. Sheldrake suggests that all matter, including the tiniest atom, possesses these morphogenetic fields, and that fields from ancient times, such as even the fields of prehistoric life forms, still exist.

Thus throwbacks occur from time to time, such as hair on a person's nose or humans with webbed feet. In a remarkable set of photographs taken by the scientist Sir Francis Galton in England during World War I, the facial characteristics of specialized groups of persons were compared. For example, Galton photographed a group of twelve Royal Air Force officers and another group of thirty enlisted men. He then carefully made composite photographs using an enlarger, one positive print, and each family of negatives for each group. Thus he created on one photo superpositions of photographs of each type; the "ideal" officer and the "ideal" enlisted man. What was surprising was that two very distinct and separate images emerged. The enlisted man had facial characteristics quite different from those of his superior.

Sheldrake argues that there is a morphogenetic field for officers that is different from that surrounding and encompassing an enlisted man. These fields are connected across time, separating the players of war games into two distinct groups. Perhaps these fields are the true causes of warfare in the first place!

I agree with Sheldrake's hypothesis. I believe that such fields exist and are beyond the normal confines of causal temporality. These fields could be nothing more than trapped qwiffs. Just as qwiffs propagate faster than or equal to light-speed and are thus able to leap the Einsteinian relativistic restraints on all objects slower than light-speed, Sheldrake's M-fields leap time's barriers.

A further consequence of M-fields is their persistence and repeatability. The eye of the newt or the regrown skin of a lacerated knee is always repeated in the same pattern. Knee skin has different characteristics from thigh skin, for example.

But what causes such persistence? In the case of a magnetic field, the bar or horseshoe magnet (the source of the field) shapes the field. Put tiny iron filings into the field and each tiny filing follows the "orders" of the master field. I would imagine that the M-field, according to Sheldrake, is somehow anchored back in the past to some prehistoric "horseshoe" that produces, say, hairy noses and armpits and balding humans. In this viewpoint the past decides what shall be in the present.

I would offer an alternative hypothesis: The past decides the present to the extent that the world is a machine. The future decides the present to the extent that the world is creative. To create something new, we need the future. To persist in something old, we need the past. The past is that aspect of living experience we label "cause." It is the persistent hope of some materialist schools of thought, including perhaps artificial-intelligence buffs, that human behavior is not creative at all and all behavior is based on prior causes, earlier events that in turn were determined by even earlier events; and so on, back to the first events that ever occurred.

Sheldrake calls the lines across time, connecting events in a causative chain, "chreodes." Chreodes are creases of time etched on the worn and weary face of earth. To create a chreode you need a principle, a point of view that acts as a guiding light. The chreodes of time are guided by nothing less than light itself.

Light is amazing on many accounts. It is the only "thing" in the universe that is its own qwiff, i.e., light is its own wave of probability as well as a physical wave carrying energy and momentum. Light moves at only one speed. Even when it appears to be moving slower as it makes its way through a material substance, it actually is interacting with the atoms composing that material and is undergoing annihilation and creation as it is absorbed and reemitted. The net effect is a slowing down in appearance of the light's wavefronts.

The guiding principle offered by light is one of economy. Light gets from here to there in the most economic way possible. It follows Hamilton's principle of least action. (William Rowen Hamilton was the Irish mathematician who is credited with the discovery of the generality of this principle to all "things" in the physical world.)

This is not the well-known and oft-trodden path of least resistance that you have no doubt heard of, although in some cases a least-action path is a path of least resistance. No, action is nature's peculiar way of keeping to a "budget." Energy is nature's way of balancing these accounts, so much energy expended for so much work done. When watching action, one watches energy and time as well as momentum and location in space.

Action involves the product of energy with time and the product of position with momentum. Action has the same character and units as Planck's magical *h* factor. To compute the action you keep track of the momentum of the object or system being observed at each step along a path, and each time you note the momentum, you also multiply it by the size of its "step" along the path that the object is moving. You also take note of the object's energy and how long it takes to cover the distances of the step taken by the object. You multiply the energy by this time interval. You then subtract this product from the momentum-space-step product and then run a tally from the starting point to the finishing point.

If you do this, and if you experiment by trying different paths — but always starting from the same point in space and time and always finishing at the same point in space and time — you will discover nature's way of keeping accounts.

For light the bookkeeping is really simple. The answer for the tally is zero. The total action used by light in going from here to there should be equal to zero, i.e., it should vanish. For light, this is true because light never speeds up or slows down. When it passes from one physical medium to another, as, for example, from air into water, the light wavefronts *are* slowed down even though the individual photons of light still move from water molecule to water molecule at the usual speed of light. Nature uses this least-action principle to actually bend the path of light when it passes from air through plastic or glass. This bending results in the common lens, of which a simple contact lens is an example. By designing the geometry of the lens appropriately we make all the least action paths of light arrive at the same point at the same time. This causes light to focus.

Particles of matter also tend to follow least action paths, but instead of zero action, particles, because they do speed up and slow down, use some action as they pass from one place to another.

Suppose that you are a particle. Let's follow your action. Again the bookkeeping is the same. Start at some point in space-time, say, at point $x = a$ and at time $t = S$. Now move one tiny space-step and observe how fast you are moving and how long it took to get there. Also observe how much energy you have either gained or lost in that movement. Multiply the momentum you have by the size of the step you have taken and then subtract the energy you have times the time it took to cover the distance you covered in the step. Now continue doing this. Perhaps on the next interval you meet with some resistance; you find that you slow down. This means that it took you longer to cover the same ground as in the previous step. Since you are on a schedule, you must reach the final point in space, say, $x = b$ at the time $t = F$. You can't dawdle.

When you do finally reach your goal, your path will have accumulated a certain amount of action. Since you were free to choose the path you took, it is possible that your path might not have been the most economical. There are probably other paths that use less action. On the other hand, you could have followed an economical action path. Then there would be other paths that were "action-hogs," using too much action to reach their final goal. The principle of least action says that if you fix the endpoints of the path — i.e., $x = a$ at $t = S$ and $x = b$ at $t = F$ — then there will exist at least one path among the many that will use the least action.

If you look at paths that are very close to the least-action path, you will find that the actions along those paths are just slightly greater than the least action. If, in attempting to consciously determine the action along a path, you do not measure the difference between the action along the least-action path and the actions along the nearby paths, these paths will, according to quantum physics, conspire. They will produce a "rut" in space-time. In moving from start ($x = a$ at $t = S$) to finish (at $x = b$ at $t = F$) along this "rut," nature appears to follow a law. This law manifests itself as the laws of physics. All laws in physics can be reduced to the one single law: Nature tends to move along least-action paths.

What happens if all paths use the same action? Then nature, not being able to discern which path to follow, follows them all. This unbridled movement can be drawn and mapped. If we look at all the possible paths from any starting point and time, but finishing anywhere at some finishing time, we draw a family of radial lines all of equal length, all focused at the starting point, and all touching the surface of a sphere at their ending points. These ending points are all equal action points, i.e., the action from start to finish is the same on all these radial paths. This is a picture of the quantum wave function.

Thus all the points that are within the same action from a given starting point form a surface or, as in the example above, a bubble. This bubble is what quantum physics predicts shall occur according to the initial condition, say, $x = a$ at $t = S$. There is no fundamental law in physics that predicts things any better than this.

But a final event, say, $x = b$ at $t = F$, does really occur. Here the laws of physics go up in a puff of smoke. There is no absolute law of predictability. The occurrence of any final event, or for that matter any event at all, can only be accounted for if it is an observed event.

If both the future event and the past event really occur, then, according to quantum physics, all possible paths between these "endpoint" events are drawn. One or more of these paths is a least-action path. Only along it is there a causal law of physics, a law of predictability. Only along it does the future point rightfully claim its heritage in the past point.

From the classical viewpoint the least-action path and the laws of classical physics say the same thing. From the quantum viewpoint neither beginning nor endpoint is predictable; there is only an infinite number of paths from any beginning point progressing to any ending point. From the quantum viewpoint with consciousness, both the beginning point and the endpoint are real because they were observed to occur. They are the limits set for the qwiff so that a path of least action can be drawn from one point to the other.

In this way consciousness creates the paths of least action and, in so doing, creates its own disappearance from the world. This vanishing act occurs because a mechanical order replaces the need for a conscious creator. Gradually, as we become more and more accustomed to time-tested and time-worn creases of habit, we fail to recognize that the future point has as much to say about the present as the past, perhaps even more.

In the next chapter we'll look at just how much control we do have over our present moments, the only moments we know.

4

Controlling the Mind: A New Vision

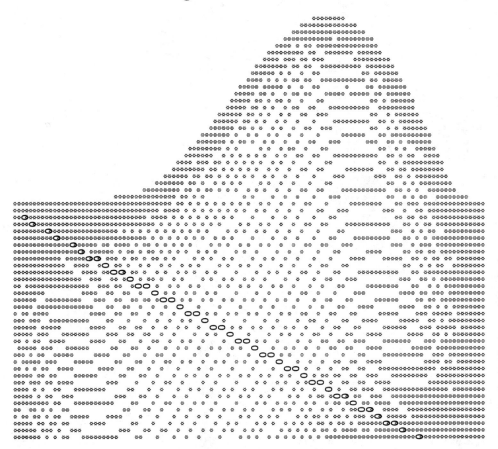

They talk of "multiple causes" and the constant "interaction" between two
"nonlinear systems". . . taking place in the dimensions of space and time.

Barry Siegel, *Los Angeles Times* staff writer (describing the difficulty that
meteorologists and oceanographers were having in explaining the origin
and persistence of El Niño, the warm eastbound current condition that
had disrupted weather patterns and wreaked havoc across three-quarters
of the globe, causing 800 deaths and $8 billion of damage.)

The Control of Nature in the "Old Physics"

One important purpose of physics is to discover and understand the forces that
govern our lives and our physical existence. The ultimate goal of some physicists is to form a Grand Unified Theory (GUT) in which there would be only
one fundamental force in the universe. Even if this idea worked, a force would
be missing from the GUT of physics. This unseen force is the force that is
always with us — the force of disorder.

The idea of force, as we presently use it, really originated with Isaac
Newton. A force is that which causes the accelerated, nonconstant motion of
any object in the universe having a mass. Thus a force produces acceleration,
or as Newton put it, "Force equals mass times acceleration." Forces of order
produce known or predictable accelerations. Thus they tell us where some
event will occur given that we are told where some earlier event has occurred.
The knowledge of "order" forces — the "forces that mold us" — provide
security and convince us that the universe is friendly.

The part of physics that deals with forces is called classical mechanics.
Mechanics is our attempt to organize, classify, and control the motion of
objects in the universe for our own benefit. Through classical, or Newtonian,
mechanics the universe appears friendly to us since we believe that a thing
mechanically controllable certainly can do us no harm. Any child or good
physicist knows that. Just watch him play with a doll or other model mechanical device.

Physicists today, using classical mechanical thinking, have classified and
thereby organized our knowledge of the force of order. At first sight there
appeared to be four possible subclassifications for the "order" force. These
were called gravitational, electromagnetic, weak, and strong forces. With the
appearance of GUT the weak and electromagnetic forces have been reclassified
into a unity — the electro-weak force. The scheme to organize the rest continues, particularly in the study of high-energy electromagnetic events involving
collisions of electrons and positrons. (Positrons are particles of antimatter
having mirrorlike symmetric properties of the electron. These events produced
the "J," or psi, particle, which is thought to support the quark theory. Quarks
carry or mediate the strong nuclear forces. The quark scheme is an attempt

at unification in the hope of grouping the electro-weak force with the strong nuclear "quark-bound" forces.)

The discovery of quantum mechanics made us see that the universe did not obey classical mechanics after all. Quantum mechanics made us realize that along with the known forces of order there existed also a fundamental force of disorder. Physicists had hoped that this force could be understood in terms of the forces of order. Some physicists attempted to model the "disorder" force by making it appear as a hidden or underworld "order" force. If such a model were successful it would be possible to control disorder at its most fundamental level. Unfortunately the model failed to lead to this result. Instead it led to a more bizarre underworld of even less control than was originally or prudently thought to exist.

Losing control is disturbing to our minds. We feel safest when we have control over the events of our lives. Our dream of control fits with our mechanical view of the universe. We hope that disorder is somehow controllable. To even think that there are forces out of our control is repugnant. For example, we all fear nuclear war far more than we fear conventional warfare. The uncontrolled release of the forces held within the nucleus puts us in a panic. Indeed, this other force in the universe, the force of disorder, is usually not thought of as a force in physics because we insist that the universe be mechanical and thereby controllable and friendly.

I have never made friends with a machine. I could imagine doing so but I believe that I would soon be disappointed. I could certainly control a machine. But after a while that would be boring. I have never been able to control a friend. Indeed, what interests me most about my friends is that they surprise me by doing the unpredictable. Similarly, the universe has many surprises in store for us through the force of disorder. Perhaps the most surprising thing about it is that this force is always with us because we are the creators of it!

Uncontrol and the "New Physics"

With the discovery of quantum mechanics in the mid – 1920s we saw that the universe did not obey classical mechanics. For example, physicists attempted to construct a theoretical model of an atom based on the known control-force of electrical attraction that exists between the atomic nucleus and its planetlike orbiting electrons. A careful calculation showed that the model failed because it predicted that the orbiting electrons would soon radiate away electromagnetic energy and thereby crash into the nucleus. This event would be catastrophic for the whole universe since all atoms would vanish. The very existence of a stable atom shows that something additional to the control-force of electrical attraction prevented it from collapsing. There had to be some "thing" keeping the electrons in orbit, without spiraling in toward the nucleus.

About 1927 Werner Heisenberg provided the missing "something." Only it wasn't a "thing." It was a *principle* that kept atoms from collapsing. He called it the indeterminism, or uncertainty, principle. It said that an object such as an electron could not be predictably known to have a defined velocity in orbit at the same time it was known to have an exact orbital location in space. And, vice versa, if one knew the electron's velocity as it moved along its orbital path, its position along that orbit would be uncertain.

Thus because of indeterminism, i.e., our uncertainty of the future, it is not possible to control an electron inside an atom. The way to see how this works is to imagine attempting to control the electron by sucking away some of its energy and thereby bringing it to a screeching halt as it spirals in toward the nucleus. Any attempt to drain its energy and slow it down will pull the electron closer to the nucleus, making a tighter orbit. A tighter orbit means a more defined position. But according to the uncertainty principle, a more defined position means a greater uncertainty in the electron's velocity. This uncertainty means that the electron now would have a higher probability of moving fast enough to elude our control and escape from the atom.

The electrons were being kept out of the nucleus by an invisible principle of human knowledge that depended on our lack of control of the future and thereby the electron's locations and velocities. It said we could not control both. To our mechanical way of thinking, there was an additional force present disrupting our control of the electrons.†

The Origin of the Unconscious: Uncontrolled Forces

Each moment is the accumulation of the infinite past and the seed of the infinite future. Yet out of this seemingly chaotic presence, there appears to emerge an order: action under our control. Our minds focus on the least-action paths (see Chapter 3), constricting visions that appear as memories. These memories are themselves chreodes, paths of least action, recorded as geometrical meanderings of neural pathways. Each branch, each curve follows a path of least action.

Something disrupts the Eden of the mind, however. The snake of indeterminism weaves its spell. It reminds us that nothing is certain. No past event is in existence. Its only record exists in our neural creases.

The future is the snake. To our "now" minds it appears as the principle

---

†One physicist who took this idea seriously in the mid-fifties was David Bohm. Bohm believed that the force of disorder could be understood in terms of the forces we believed were controllable. He attempted to model the force of disorder by making it appear as an orderly but hidden force-field, like a "strong" nuclear force-field. Through the success of such a model it was hoped to control subatomic events. Although the model was successful (it was possible to create a mathematical expression for the force of disorder acting on the atomic electron), it unfortunately failed to lead to this result. Like the proverbial joke, the operation was a success but the patient died. Instead the model led to a more bizarre unruliness wherein forces exerted on electrons at one place would *instantly* disturb electrons at another location, in violation of the very mechanical decorum the model was trying to preserve.

of uncertainty. It beckons to us with forebodings and promises and appears in our dreams.

But why does it come? What does the future need or want from the present? The answer is manifestation. The future reaches back to the present and "shakes hands" with it in agreement that between them, and only between them, shall there be a path of least action. To our minds in the present moment, this ghostly apparition appears frightening or perhaps reassuring.

Think of the future as a series of pictures of yourself. Each picture portrays a different you, much as an actor changes his guise with makeup and mannerism. Each of these futures lives in the time ahead of us just as our neighbors live in the space around us. Each friendly "neighbor" calls to us, perhaps over the backyard fence or on the telephone, and invites us over. We must choose whom to visit.

In the same sense our futures call to us. Some "neighbors" appeal to us more than others, so we are more likely to go toward their home. Some futures are more inviting than others, so we shake hands across the "backyards of time," agreeing to get it "together." But there is competition for our attention. Several "neighbors" want us at the same time. From the standpoint of our present, these choices make our futures uncertain.

Now think of the past as another series of pictures of yourself. "Ah," you say, "that's not possible because the past is real." But how real is it? Freud claimed that our memories were filled with pasts that we didn't even know existed. Jung felt that all humans held archetypal memories constituting a collective human unconscious. Ask any witness to describe events that were witnessed by others and most often you find a lack of corroboration. It's as if the witnesses saw different events that were somehow similar rather than the same event from different viewpoints.

In a now classic paper, Einstein and Tolman showed that if quantum mechanics describes events, then even the past is as uncertain as the future.

Thus, how do we have any pasts at all? The answer is that we create them! Now I know how bizarre that sounds (after all, I am a physicist). This would mean that we in the present are responsible for our pasts, not the other way around. We are the creators of history. What we call the past only exists in the windmills of our mind, the meandering chreodes of our neural systems. Even the words we write, as indelible as they appear, as if etched in the consecrated halls of concrete, are just the words of men and women. Words spoken from the context of memories constantly bombarded by the vagaries and uncertainties of Heisenberg's principle and by beckonings of the future.

What we call our unconscious is nothing more than the uncertainty principle at work on our neurons. Our dreams and our unspoken fears result from the conflict between our future and our past.

Just as the future beckons for our attention, so does our past. It looks toward us in the present and asks for agreement. Like a child left by her parents, the child that is our past cries for us not to forget her, to agree that we, the present, are still in control.

To understand how each of us creates an unconscious we need to look back for a moment at Chapter 1. There we looked at how the complementary aspects of our observations alter the physical contents of our observations. Specifically I showed how a particle in a tube was affected by attempts to observe it either in a specific energy state, i.e., moving fast or slow, or at a particular position in the tube, on the right or left side. If you now think of the tube as a neuron or as a molecular chain in a cell or as a particular molecule, such as the amine group, acting within a neuron controlling the release of neural transmitter in the synapses, you will have some idea of the quantum mechanics of our nervous systems.

At this stage of research it is difficult to speculate much further than this. Certainly our thoughts alter our physical bodies, much as an amoeba alters its body in order to move from one place to another. It matters little where the dividing line exists that separates the macroscopic mechanical operational systems of our bodies from the microscopic quantum mechanical "minichip" systems creating thought. At some time in the near future scientists will discover the primary quantum-mechanical "chip" of the nervous system. The important hypothesis is that such a "chip" or "tube," as I have modeled it, exists. The particle that acts as the basis for behavior modification may be an electron in a complex molecule or a molecule itself. Whatever that "particle" is, wherever it exists, it is the complementarity of the ways in which it can be observed that determines our destinies.

Now, when an observation takes place, it makes the complementary observation "spread out" in Hilbert space, the abstract space of our minds. Thus, for example, a position observation in the tube makes the energy observations uncertain. They are, in a sense, "pushed" below the horizon of consciousness.

I am much amused by the terminology of modern computer science. There one finds the operational terms *push* and *pop*. A push operation occurs when a stack of data, much like a stack of dishes in an automated restaurant, is pushed down to accommodate a new "dish" of data at the top of the stack. A pop is the opposite action, the removal of the top "plate" of data to be sent to a register or to another part of the computer system.

In a similar manner the complementarity of human observation "pushes" and "pops" the data of our observations, only a "pop" of position is simultaneously a "push" of energy, and a "pop" of energy determination is a "push" down of the complementary measurement. Each "pop" is accompanied by a complementary "push." There cannot be a "pop" without a complementary "push." This is the "new physics" law of action and reaction: For every quantum "pop" there must be a quantum "push."

According to Freud we all possess defense mechanisms whereby any unacceptable impulse or idea is rendered unconscious. This mechanism is called repression. But how does this mechanism work? Why is it sometimes

111

easy to recall unpleasant memories, i.e., repressed thoughts, and at other times nearly impossible? Also, when a thought is rendered unconscious, pushed down below consciousness, what happens to it? Does it leave the vicinity of the ego and bury itself in the id?

Using quantum physics as our basis, it is the operations performed by our observables that repress unpleasant thoughts. Since each observation requires a quantum-mechanical operator to operate on the qwiff, or contents comprising our id, the result of each observation is the "removal" from consciousness of the results of a complementary observation. For example, the observation of a series of neuronal "tubes" can be performed in many ways. You could choose to observe only positions in each tube, or just the position of every other tube, or every third tube, etc. With each position observation the energy state of the tube is "repressed." On the other hand, an energy observation in a tube represses knowledge about the location of the particle in the tube.

A particular sequence of position observations, Pr and Pl (position right and left, respectively), and energy observations, Mf and Ms (motion fast and motion slow, respectively), could appear as $Pl-Mf-Pr-Mf-Pr-Ms-Pl-Mf-Pl-Ms-, \ldots$, assuming that the sequence has alternate position- and energy-determination operations (observations) occurring. It needn't matter whether the result turned out f or s for an energy observation or r or l for a position observation. The memory is contained in the alternate sequencing of the P and M complementary observables.

By repeating the same sequence of operations over and over again, the results of the observations will take on a random order. That is, the lowercase letters, $f, s,$ and $r, l,$ will be randomly distributed or mixed along the neural network pathway. However, the sequence of P's and M's will not be disturbed. In this way a habitual thought pattern is created and at the same time the complementary pattern that could have resulted from the sequence starting with an M observation — i.e., $Ms-Pr-Ms-Pl-Mf-Pl-, \ldots,$ — is repressed.

Now the information in each "tube" is still present, but to reach it one must determine the sequence of the repressed set of observations. After only one repeat, or when the thought first occurs, the qwiffs in each tube have not been "mixed." They each retain some of their purity. When the attempt to repeat the repressing sequence occurs, the qwiffs in the tubes become mixed.

A mixed qwiff results from repeated operations of sequences of observables on different tube sets. For example, suppose there are five tube sets, each consisting of ten tubes. A particular pattern might appear (see Figure 26A).

Each row of the table in Figure 26A contains the results of repeated and, therefore, "learned" operational procedures. The pattern of P's and M's, appearing in an alternating sequence, is arbitrary. The results show qwiffs in the tubes. Thus l means that the particle in the tube is on the left side of the tube and that the qwiff in the tube appears as shown in Figure 13 in Chapter 1. Similarly for the other letters. If we were to assign arbitrary numerical values for $r, l, f,$ and $s,$ we could determine an average value for each column

QWIFFS IN TUBES, OR HOW TO REMEMBER TO FORGET

```
observation
sequence-----> P   M   P   M   P   M   P   M   P   M
              +--+---+--+---+--+---+--+---+--+---+--+
                   f       f       f
set 1 ----->   r  |  1 |   1 |       r  |    r  |
                                      s           f
              --- --- --- --- --- --- --- --- --- ---
                           f
set 2 ----->   r     r     r     r     1
                  s           s       s       s
              --- --- --- --- --- --- --- --- --- ---
                           f       f
set 3 ----->   1     1     r     1     r
                  s           s
              --- --- --- --- --- --- --- --- --- ---
                                       f
set 4 ----->   1     1     r     1     r
                  s     s     s               s
              --- --- --- --- --- --- --- --- --- ---
                   f       f       f
set 5 ----->   1     r     r     1     r
                              s           s
              +--+---+--+---+--+---+--+---+--+---+--+
```

Figure 26A

in the table. Suppose that we take:

$$r = 0 \quad l = 1 \quad f = 2 \quad \text{and} \quad s = 3$$

For example, the average value for column one is .6. This means that the qwiff pattern is 60 percent a left pattern and 40 percent a right pattern. Similarly for column four the average is 2.2. This means that the qwiff pattern (just look at the decimal fraction) is 20 percent a fast-moving pattern and 80 percent a slow-moving one.

These averages represent what physicists call a mixed state. The qwiff is never actually in a mixed state. It is always pure and follows the rules of operator-observables, thus conforming to the boundary conditions set up by observation. But the average qwiff does have significance. It means that the next time a similar measurement is carried out, i.e., the above sequence of P's and M's is repeated, the resulting pattern for that row of ten tubes should produce results that differ little from the average qwiff. Now in this case this is not too difficult to achieve because the values chosen were well separated. A value of less than one is always taken as a position determination, while a value greater than two is always taken as an energy determination.

But since each qwiff is a linear combination of the complementary qwiffs[†] it would be more accurate to represent the observables by:

$$r = 0 \quad l = 1 \quad f = .707 \quad \text{and } s = -.707$$

The value .707 is the square root of .5. If we perform an average for each column, the results will begin to overlap. For example, for column one we find once again .6, but for column four the average is $-.1414$. This value is close to zero, the expected value for a position measurement on the right, but it is an energy measurement saying that the qwiff is slightly more likely to be a slow-moving pattern than a fast-moving one.

It is only by continual repetition and thus adding more and more tube sets to the ensemble composing the average that the respective tubes in their sequence begin to show well-separated averages. Thus when another measurement is performed, its particular numerical sequence will appear to the ensemble average as "just one of the boys," that is, merely a repeat of what has gone on before.

In other words, as the ensemble size increases, we find for the expected result the value one-half for position measurements (the P columns) and the value zero for energy measurements (the M column) simply because, after many measurements, these values are the normal values for those measurements.

With a larger and larger ensemble the average result actually measured approaches the predicted result more and more closely. (Note that the predicted result for a P, or position, measurement is simply one-half, zero plus one divided by two, just the average for the two possibilities, the same as for a two-sided coin. A similar situation occurs for an M measurement, only the result is zero.)

Now it is important to remember that nothing is held down and fixed in memory. We are dealing with qwiffs in tubes, not bits in chips. A chip bit is either zero or one. It doesn't care whether a P or an M operation is carried out on it. Although constituting a bit of memory that is held down and fixed, it has no memory; but a qwiff that is capable of wiping out or repressing memory has a memory. The way the qwiff memory works follows from the principle of superposition discussed in Chapter 1 (also see the footnote below), an r qwiff is composed of a superposition of an f and an s qwiff. The same is true for an l qwiff. Similarly an f qwiff is composed of a superposition of an r and an l qwiff. In other words, each qwiff is composed of "unconscious" manifestations of the complementary properties. An r qwiff has repressed within it "memories" of fast and slow movements. To rekindle the repressed thought (or thought bit) an M operation must be performed on the tube containing the r qwiff. This will "pop" the qwiff and alter the memory.

---

†In Chapter 1 (see Figures 8, 9, and 10) we saw that qwiffs can be combined by adding or subtracting. Thus, $f = (r + 1) \times .707$ and $s = (r - 1) \times .707$. This leads, by simple algebra, to $r = (f + s) \times .707$ and $l = (f - s) \times .707$.

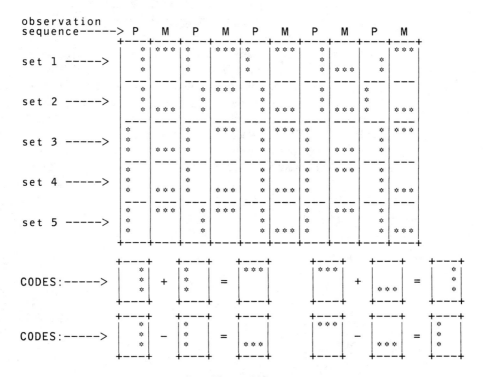

Figure 26B

A picture of repression: The asterisks (*) depict the results of performing P and M measurements. The codes show how P states can be added or subtracted to produce an M state, and vice versa. Thus a P state is a repressed superposition of M states, and vice versa.

Suppose that you are trying to remember a repressed thought. Somehow you return to the neuronal tube sets involved with the pattern set of the particular sequence of *P* and *M* operations that you used at an earlier time to "forget" that awful past event. You try to remember but nothing happens. Why not?

Be patient, something has happened. Perhaps one set of tubes of your neuronal ensemble has been disturbed. Instead of repeating the sequence of, for example, *P–M–P– M –P–M–*, . . . , you follow a different pattern, perhaps *M–P–P–M–P–M –M–*, . . . , randomly attempting to rekindle the tube sets. Each new pattern upsets the established pattern. The results begin to produce values significantly different from the tube average. Flashes or recalls are occurring. You try again by, perhaps, not trying at all. This results in a reversal of the fixed pattern. A particular tube set of experiences, *M–P–M–P – M –P–*, . . . , instead of *P–M – P–M–*, ... The flash becomes stronger.

Suddenly the new pattern is repeated throughout the ensemble. The averages are changes in each column of the table shown in the previous figure. You remember what you had repressed.

I believe that this scenario, or something similar to it, is played out when a repressed thought is made conscious again. I also think that the difference between a preconscious and an unconscious repression is simply the difference in the size of the neuronal ensemble involved in the "act of repression." The preconscious differs from the unconscious only in the number of neuronal events involved in the act. When we deeply repress, we push down a data stack of qwiffs by performing a "push" operation over and over again. A preconscious "push down" is not repeated as many times. It is tolerable that we have that pain.

The origin of the unconscious is the quantum-physical nature of our neurons. We cannot help but to forget in order to learn to remember.

5

Order, Disorder, and Observation

The transducer — the machine, as
instrument and as message — thus
suggests the sort of duality which
is so dear to the physicist, and is
exemplified by duality between wave and
particle. Again it suggests the
... bon mot — I do not remember
whether it was Bernard Shaw's or Samuel
Butler's — that a hen is merely an
egg's way of making another egg.

Thus the machine may generate
the message, and the message may
generate another machine.

This is an idea with which I
have toyed before — that it is
conceptually possible for a human
being to be sent over a telegraph line.

Norbert Weiner
God & Golem, Inc.

Order and Mind

It happens in every movie in which a criminal is brought to justice. As he approaches the bench of the courtroom, the judge shouts out to the buzzing gallery of onlookers and sensation seekers, "Order, order in the court!" The people, not really knowing what the judge wants, simply shut up. A few of the onlookers interpret the judge's remarks to mean pay attention and shut up. So they turn to the front of the room and stare intently at the judge.

What does the judge want? He cries for something called order. But what is order? Is it enough that the spectators "come to order" just by shutting up and facing forward?

In the military, officers give orders to enlisted persons. How are these commands to perform related to order?

In this chapter we will look at the concept of order as it relates to the mind, and specifically how this concept has been altered by the principle of uncertainty in quantum physics. This will prove to have an important bearing on what we mean by knowledge and information, and that, in turn, has relevance for our model of the mind.

The concept of order in physics turned out to have a relationship to the behavior of large numbers of things moving or existing as an ensemble. In physics such ensembles have average properties. The relationships that physicists draw between these average properties constitute a field of physics called thermodynamics and, when more detail concerning the distributions associ-

118

ated with these averages is required, the field of statistical mechanics. Each of these fields belongs to the realm of classical physics, where our story starts. As you read about order, disorder, and the term *entropy,* the name given to a certain mathematical function that provides a measure of the disorder in any system composed of many objects, keep in mind our judge crying for order and the dilemma of the military officers commanding a field of enlisted persons.

Disorder and Information

Even though classical physics no longer represents the way we presently understand the physical world, it provides our first version of order and disorder. The part of classical physics dealing with order on a microscopic scale is called statistical mechanics. It applies to classically defined objects — those that have well-defined positions and velocities at each instant but which are too numerous to practically count. Under normal circumstances even atoms and molecules can appear to be "classical." The molecules of a gas in a chamber enclosed by a piston provide a typical example. Under most circumstances these molecules behave as if they were perfectly hard little spheres that only interact by bouncing against each other and the walls that confine them. This behavior is known as the state of a classically perfect gas.

Recently a new and surprising relationship was discovered between the state of such a gas and, a subject that has enormous implications for us today, *information theory.* In this chapter we will also explore this relationship and the more general association it has with the human mind, consciousness, and the "new physics."

Science has discovered that any time there is a change in the environment surrounding a system, information is actually imparted or extracted. For example, let's look at our perfect gas again. By suddenly pushing on the piston, thereby confining the gas to a smaller volume and also extracting enough heat from the gas to keep it at the same temperature, we actually gain information about each molecule. The reason is that we now know with greater probability the location of each gas molecule because each would be confined in a smaller volume of space. This knowledge implies that the gas would be more ordered than before.

Physicists use the term *entropy* as a measure of the amount of disorder "contained" within the enclosed gas. In this example, by maintaining the temperature of the gas and decreasing its volume, the entropy, or disorder, has decreased. In principle it would be possible to squeeze the gas so that no space existed between the molecules (like stuffing Ping-Pong balls into a large box) and, according to classical physics, it would always be possible to extract enough heat from the gas as it was compressed so that a state of complete, but rather stuffy, order would be obtained.

In contrast, if we were to squeeze the gas without extracting heat from it, we would find that the amount of disorder in the gas was exactly the same after as before the push of the piston. This occurs because, although we would have gained information about the locations of the molecules — since they would now be contained in a smaller volume of space — we would have lost information about their movements. Just as a slow-moving Ping-Pong ball is whacked by a rapidly approaching paddle, each molecule would be randomly struck, and therefore move faster, as a result of hitting the piston. Thus the gas, composed now of faster-moving molecules, would actually get hotter.

With further attempts to confine the gas, eventually quantum physics becomes important. No matter how cool we make the chamber as we compress the gas, we would find that we could not obtain total order. Greater confinement of each molecule would produce, according to the uncertainty principle, a greater uncertainty in its possible speed and therefore less certainty about its individual behavior. The gas would exhibit what is called zero-point energy. Even though its temperature was reduced to absolute zero, the molecules would still continue to move. Each molecule, however, would no longer be able to occupy a single position at a single time. Instead each would "spread out" throughout the whole volume of the chamber.

This phenomenon is called Einstein-Bose condensation and is observed in low-temperature helium. In this "condensed" state the molecules do not have individual existences. Instead they behave as if each molecule was a fluid filling the whole volume.† This resistance to perfect order that takes place at these low temperatures enables a new state of order in matter, called a superfluid, to occur. Quantum mechanics disables our attempt to have perfect individual control. We reach a new collective experience that cannot be predicted from classical statistical mechanics. For example, in this superfluid state the gas exhibits no resistance to the movement of an object through it. It has absolutely no viscosity. A similar situation occurs with certain supercooled metals called superconductors. These show remarkable electrical and magnetic properties. An electrical current, once started in a superconductor, continues without resistance. A magnetic field "caught" by the superconductor is totally enclosed by it. No magnetic-field lines leak out.

Order, Knowledge, and Maxwell's Demon

"Super" behavior results from quantum mechanics "taking over" the behavior of classical objects. It also alters the meaning of information. It does this by connecting the concept of order with our concept of knowledge. To see this,

---

†As a result of the uncertainty principle a well-confined particle will begin to "zero point" oscillate using the limits of space put on it to "create" energy. This means that the well-defined confinement of a particle in a cooled gas or liquid begins to make the gas or liquid boil even though the temperature is low. This "boiling point" is given the special name of *condensation temperature* and was first discovered by Albert Einstein. This means that if the temperature is lowered still further past the condensation point, the gas will begin to behave in a bizarre manner in which the individual particles lose their separate identities. They cannot be said to be occupying such small spaces any longer. Instead each particle fills the whole volume!

imagine a tiny molecular-size demon sitting inside a box of gas.

This demon was conceived by the fertile mind of James Clerk Maxwell during his theoretical heyday in pre-Victorian England. Maxwell, you may remember, was the creator of a set of equations that described the electromagnetic field and predicted electromagnetic waves, resulting eventually in the discovery of radio and television.

In 1871 Maxwell wrote *The Theory of Heat.* In this treatise he describes a fictitious character, "a being whose faculties are so sharpened that he can follow every molecule in his course, and would be able to do what is at present impossible to us . . . ." This being is called Maxwell's demon and usually appears in physics classrooms when the subject of thermodynamics or statistical mechanics is discussed.

Today we are facing a microelectronic revolution. What was in Maxwell's day impossible is on the threshold of realization in our time. In 1871, however, the "demon" was a pure fantasy. He was created because Maxwell wondered about how the behavior of a hot gas depended on the movements of its molecules. He realized that a gas got hotter if the molecules moved faster. He also came to the conclusion that the molecules were not all moving at the same speed. As they bumped into each other, some would undoubtedly be slowed down at the expense of the others speeding up. This would result in a distribution of speeds in the gas and, as we saw in Chapter 3, an average speed for the molecules. As you might not expect, the distribution turned out to be none other than the "normal" distribution we looked at in Chapter 3. Instead of the number of occurrences being the distribution variable, the speed of a molecule is now distributed, with most molecules within one standard deviation of the average speed.

Maxwell showed that the average speed changed with temperature. If the temperature quadrupled, the average speed doubled. The temperature depended on the kinetic energy of the average molecule. If the average molecule's kinetic energy (this is $mv^2/2$ — one-half times the mass of the molecule times the square of the velocity, or speed, of the molecule) changed, the temperature of the gas would change correspondingly.

Maxwell then thought up the demon. He imagined an intelligent little character sitting in a box filled with gas. Let the box be divided into two and suppose that there is a small slide door in the partition separating the compartments. The demon operates the slide to select fast molecules from slow ones. He hopes to fool the physicists watching the gas by collecting the fast molecules in one compartment and the slow ones in the other. Thus he is attempting to beat the odds, which predict that each compartment will have an equal number of fast and slow molecules, by using nothing more than his ability to discern fast from slow. In this way he will cool down one side of the box and heat up the other. He will create more order in the cool chamber and more disorder in the hot one. At first he has no trouble. Even at room temperature the gas is hot enough to have quite a bit of variability in molecular speeds. His only concern is his ability to move the slide door quickly enough so that only one particle is able to pass through the partition at a single time. Being clever, he makes the door hole as small as he likes, but not too small to block out any molecules from passing through. By making the doorway sufficiently tiny he has more than enough time to work the slide.

He does have one problem, however. He cannot make the doorway too small or else the uncertainty principle begins to take its toll. If the doorway is made too tiny, it creates an uncertainty in the speed of the particle passing through. Thus the demon must make a trade-off. He must compromise the size of the door and the time it takes him to operate the slide.

On the other hand, if he makes the doorway too large, he finds so many particles passing through each time period that he becomes unable to tell just when a particle passes through. If he, in spite of the large doorway, does note the time of passing of each particle, he again faces the uncertainty principle because each time determination introduces an error in the energy of the particle as it passes through the door. This error wipes out the very discernment he is seeking to make, fast from slow.

Suppose that by "fast" we mean particles moving with speeds in excess of the average speed and by "slow" those that move with less than average speed. Now, our demon is very human when it comes to observation. He must be able to discern fast from slow and he has only so much ability to do so. To make such a discernment requires the ability to "split hairs." The greater the ability, the tinier is the difference between the speeds the demon is able to detect.

If we label the tiniest difference between speeds that the demon can resolve by his discernment Δv, we can determine just how tiny Δv can be.

If you remember the bell-shaped curve of Chapter 3 for a moment, you will see his predicament. The curve has a natural width, i.e., one standard deviation. He must be sharper than that, i.e., his Δv must be narrow enough, to see if a particle is above or below average speed. If his ability to recognize speed is blurry, that is, if his Δv is too wide, and the blur spreads beyond the width of the curve, he won't be able to tell fast from slow. This is analogous to attempting to paint two fine lines on a canvas. If the brush width is wider than the two lines to be painted, the brush stroke will wipe out the difference between the lines. The demon's ability to measure speed, his Δv, and thus come to know fast from slow, is limited by how narrow the distribution of speeds is.[†]

[†]As the temperature drops, the spread in the range of speeds also narrows. We saw this occur in Chapter 3 as a result of our becoming more and more certain of the event under examination. Thus, below some temperature our poor demon will not be able to tell fast from slow because all the particles will be moving at speeds very close to each other. Or, put another way, the demon's ability to distinguish must be fine enough to see fine differences in the velocity or energy spread of the particles in the gas. The spread in the speeds of the gas acts as an upper limit to the demon's ability to tell fast from slow. So far there is no problem. But there is also a lower limit on the demon's ability. Remember that he is using his intelligence to heat selectively one

So our demon's ability is caught between two limits. He must be discerning enough to tell fast from slow, but if this discernment is too fine, he begins to alter the speeds of the particles in uncontrollable ways introduced by the principle of uncertainty. Thus we are at the frontier of knowledge, the line between what can be known and what must remain unknowable unless the very act of attempting to know disrupts the very order one is trying to create by observational operations in the first place.

Why the World Is Probable and Not Actual

The limits of knowledge pointed out by the demon's endeavors are soon to become the concern of the microprocessor computer industry. Already we are facing the advent of the molecular microchip. Questions raised by the demon will resound in the laboratories of the computer industry.

I believe that the demon is with us in our everyday thinking processes. Discernment is a question of trade-off. In the above example, if the gas remained at room temperature and the number of particles contained in the box was not too great, the demon would have little trouble in operating the slide. Under these classical conditions, the little fellow, if he existed, would create quite a stir for his observers. They would gradually see one side of the box get hot while the other would grow cold. Such behavior is not consistent with the second law of thermodynamics. This law forbids heat energy from flowing from a cold body to a hot body without someone doing some work to make that happen. The demon is not doing any work and yet the hot side of the box grows hotter at the expense of the cold side. Now this violation of the second law is not a violation of the first law of thermodynamics, which states that energy involved in any thermodynamic process must be conserved. The heat that flows from cold to hot makes the hot side hotter by making the cool side colder, thus conserving energy.

The second law is a law of probability. The demon is able, provided the gas is classical, to violate the second law. If the conditions surrounding the demon become quantumlike, i.e., the gas temperature becomes too low or the number of particles in the box is too large, the demon's operations become very disruptive. It is the classical environment that enables the demon to do any observations at all.

In *Science and Information Theory,* Leon Brillouin, the French physicist, adds a new idea to the demon's repertoire. The demon is not actually violating the second law of thermodynamics because he is increasing the entropy (the amount of disorder) of the whole system, including himself. He does this because he must make an observation in order to tell fast from slow. And, as I pointed out in Chapter 2, every observation requires an interaction. For the

side of the box while at the same time cooling the other. This lower limit arises because of the uncertainty principle.

demon to "see" the particles in the gas he must use some form of light. That light must be high enough in energy to stand out beyond the thermal energy background of the heat (infrared light) produced by the particles themselves as they interact with each other. Thus the demon needs a torchlight that produces photons (particles of light), each with an energy greater than the background thermal "noise." Taking this into account we find that the interaction of the demon's light with each gas particle creates an increase in the entropy because light energy passing from the torch to the gas particle follows the second law of thermodynamics.

When the torchlight's photon returns to the demon's eye, another increase in entropy occurs. When the information registers in the mind of the demon that the fast particle has entered the hot side of the box, thus creating more order in the system, there is an entropy decrease. Taking all the entropy changes together, including the demon and the gas, we find the system's entropy higher than it was before the demon began his observations.

Thus it appears that "we cannot get anything for nothing, not even an observation," as aptly stated by physicist Dennis Gabor, the inventor of the hologram. The world grows more probable and chaotic with each observation. And therein lies the hope of mind and humankind.

Each observation creates a knowledge of order and a pocket of organization in the universe at the expense of more disorder in the whole universal space-time fabric. The "fuel" for that observation is time. The march of time is driven by the drum bands of entropy. It is the ultimate fact of entropy increase that is the background for order in the universe. Our successes and failures can only occur if the entropy of observation is irreversible. What happened yesterday cannot determine the results of today. The temporal sequence of one miracle after another, and all observations that create order out of the ever-increasing disorder are miraculous, means that nothing is ever won without effort. That effort results in the entropy production, which is time's march to eternity.

How that effort is manifested is another thing. Perhaps by forgetting or becoming confused it would be possible to pay the entropy price of allowing the physical universe outside the mind to manifest "sudden" order. Perhaps this was God's trick. He made the big bang of creation when She forgot She was He too, i.e., She lost sight of Her own Godliness.

Each observation turns a probability into an actuality creating bits of knowledge, pockets of order, and fabrics of disorder. When observation ceases, according to quantum physics, the universal qwiff becomes purified with no further increase in entropy. Observation is the cause of all disorder.

124

The Act of Observation: How Actuality Arises from Probability

Finally we come to it. What is the universe and how does it come into existence? Well, the universe is a kind of machine. Our machines, the ones humans build, are "models" of the universe's order/disorder manufacturing capability. The second law of thermodynamics, which has direct bearing on the concept of information and order/disorder, was first discovered by S. Carnot in the early nineteenth century when he realized that any machine that converts heat into work must always transfer some of that heat to a "reservoir," where it becomes unavailable to do work. That reservoir is always at a lower temperature itself than the source of the heat. Consequently, heat will never spontaneously flow from a colder body to a hotter one. Some years later R. Clausius put this idea into the concept called entropy. All physical processes involved in mechanical heat conversion lose some of the heat to the background, never to be regained again. It is this unavailability of heat energy that is the effect of the law of entropy increase. The price of energy conversion is entropy production.

Conversely, to transfer heat from a cold body to a hotter body, work energy must be imparted to the system. The body that is cooled off does lower its entropy, but the rest of the universe pays the price.

By the end of the nineteenth century Ludwig Boltzmann realized that entropy depended on the number of possibilities available to a physical system. The greater that number, the greater the entropy. Thus, if in the normal passage of events a system "evolved" to where it contained more possibilities than it had initially, the entropy of that system had to increase.

And that was the universe's game. The mission of the universe is to seek out all possibilities, be copious, giving all things equal opportunity to "do their thing." Our small pocket of the whole picture, our earth and its life forms, are a result of that grand plan.

But why bother? I mean, why should the universe care to follow such a plan? The answer seems to be, in order to become conscious. Consciousness necessarily demands events, real occurrences in space and time. Events, according to the ideas presented in this book, can only occur if they are recorded in the mind. These events are the registrations in consciousness we call observations.

But, as we came to realize through Brillouin's work, mentioned earlier, such occurrences, although they themselves result in order arising out of disorder because mind has gained knowledge, require interactions that involve an overall entropy increase, i.e., an increase in disorder. If, no matter what we do, the universe produces more and more disorder, then how does any order take place?

It is here that quantum physics comes to the rescue. Order is created in the correlative behavior of qwiffs. When two objects interact and are not observed, their qwiffs become entangled and inseparable. Instead of maintaining separate qwiffs, each object joins a single qwiff. This results in a cooperative behavior between the objects even though they may no longer be in interaction with each other. This kind of cooperation plays a vital role in life and is necessary for the existence of all chemical activity.

Through such correlations, where independent objects join through mutual interaction into one qwiff, entropy is decreased. Cooperative behavior creates order, and human thought plays a major role in that enterprise.

Order, Entropy, and Human Thought

"Entropy" is a human thought. So is "universe," "qwiff," "quantum physics," and every other word that exists in every language that is spoken or written. Thought is an ordering experience for the thinking machines that carry it out. Just as our machines are emulations of the universe's grand entropy-production plan, our minds are emulations of the universe's grand entropy-reduction plan. Our thoughts are representative of the universe's thoughts. Thoughts produce order in the thinker.

They can also produce violence and disorder. Thoughts can destroy the thinking machines. In an amusing "Star Trek" episode Captain Kirk, in order to save his ship, the *Enterprise,* has to talk a destructive computer device that has come aboard out of blowing up the ship and all "carbon units" inside. Kirk, using a logical paradox, convinces the machine that it must destroy itself because it has made a mistake in its "prime directive." The machine turns its destructive power onto itself. The crew of the *Enterprise* beams the device into space in the nick of time, thus saving the ship when the device self-destructs.

In a similar manner we are each able to produce self-destruct thoughts. Indeed we seem to have a great talent for it. In this section of the chapter I want to explore the two types of thought humans are capable of. I wish to offer this possibility: Thoughts that are correlative, involving synthesis and analogy, create order and reduce tension, produce a sense of well-being and a healthy lack of destructive ego tendencies. On the other hand, thoughts that are noncorrelative produce qwiff "pops" that result in entropy production in the thinker, causing confusion, senses of loneliness, powerlessness, and fear.

In Claude Shannon and Warren Weaver's little book, *The Mathematical Theory of Communication,* Weaver quotes Sir Arthur Eddington, the prominent English physicist and cosmologist, who said, "The law that entropy always increases — the second law of thermodynamics — holds, I think, the supreme position among the laws of Nature." Shannon's recognition that entropy not only means randomness and chaos, disorder and noise, but also information was a stroke of profound genius. This stroke, as we shall see later in this book, has significantly affected the roots of quantum physics as well.

We see that the universe "knows" the wisdom of entropy — produce and give to all things their equal chance to manifest. And therein lies the crux. If all phenomena have equal chances of success, why do some things emerge triumphant, evolving over others? How does nature weight the cosmic crap game in favor of a few shakers and rollers?

With all possibilities being equal, entropy emerges the victor. As soon as some possibilities disappear, that is, become impossible, entropy is forced to retreat.

The mystery is solved as soon as a mechanism is introduced wherein one object can interact with another by exchanging something, as, for example, in the interaction between electrons in atoms exchanging photons of light. Through interaction previously independent qwiffs become interdependent, and that means more order and less uncertainty, less available information. Entropy is reduced.

Each interaction and each interchange is the universe's cry of loneliness answered. Objects that are independent of each other, taken together, possess more information and therefore more entropy than objects that are interacting or have interacted. For example, suppose there are two electrons in space. Each electron possesses a certain property called *spin*. This can be thought of as a rotational movement like the earth's rotation or the spin of a baseball as it is thrown by a pitcher in a baseball game.

This spin has a direction associated with it. If the electron spins clockwise, we say, for example, that it has "spin up." If it is rotating counterclockwise, we say that it has "spin down." Whenever an electron's spin is observed, it is always discovered in either an up or down spin state, much as a flipped coin is found, once it lands, as either heads up or heads down. The information associated with the electron's spin depends on the number of possible observable states. For one electron there are, therefore, two states, U (for up) and D (for down). If there are two electrons independent of each other, there are four possible states: UU, UD, DU, and DD. With three independent electrons there are eight states possible: UUU, UUD, UDU, UDD, DUU, DUD, DDU, and DDD. Each time you add another electron to the mix, the number of states increases by a multiplicative factor of two. Thus with four electrons there are sixteen states, five electrons produce thirty-two states, etc.

Returning to just two electrons for a moment, it is possible to calculate the information associated with their independent existences. The formula for this calculation depends on a certain mathematical function called the logarithm. Briefly, the logarithm of a number is the answer to the following question: What is the power to which you must raise a base number in order to arrive at a number given or specified? In the case of the electron spin situation, we let the base number be two, corresponding to its two possibilities available, U and D. The number specified or given is the total number of states possible. Thus, for example, with two electrons there are four states. So we find that the logarithm (to base two) of four is the power we must raise the base two in order to reach four. Since two times two equals four, or four is two raised to the power two, the logarithm is simply two, or in shorthand notation, $\log(4) = 2$. With three electrons, we find eight states, so we seek $\log(8)$ and find that $\log(8) = 3$ because two raised to the power three is eight. With sixteen states, $\log(16) = 4$, etc. In the simple case of the two-state electron spin, the logarithm is just the number of electrons present.

According to Shannon's work on information theory, it is this log function that determines the information's measure. According to Boltzmann's work in statistical mechanics, it is the same log function that determines the entropy. That is why entropy and information are really the same thing.

Again returning to our example of two electrons, since there are two states possible for each electron, each one has an intrinsic information, $I = \log(2)$. Taken together, and maintaining their separate but equal existences, we simply add their informations to get the total information available. This is

$$I(\text{total}) = I(1) + I(2) = \log(2) + \log(2) = 1 + 1 = 2$$

Simple enough. With three electrons acting independently we find

$$I(\text{total}) = I(1) + I(2) + I(3) = 3 \times \log(2) = 3$$

and so on. With four electrons, $I(\text{total})$ is four.

Since each state is equally possible, the entropy, and therefore the information available, is maximum. All possibilities get equal chances.

But now comes the means by which entropy is reduced. The only way this can happen is through quantum physics. According to its laws, whenever the two electrons interact, their qwiffs correlate, the behavior of one electron affects the behavior of the other. From this correlation a new state of two electrons emerges triumphant. In this state not all possibilities are equally likely.

For example, two electrons may enter into a total, or correlated, spin state of zero. In order to create this state, nature — in accordance with its own "nature" — demands that all possible ways that both electrons can yield a total spin of zero be present. This means that there will be uncertainty as to which electron has which spin. In this state all we know for sure, according to the uncertainty principle, is that if one electron has spin up the other must have spin down. What we don't know is which has which. The qwiff specifying this peculiar situation cannot be written as UD because UD would mean that we know that the first electron (electron one) was U and the second was D. Neither can it be written as DU, and for the same reason, namely, that we would have to know which electron had which spin. Either way we would know more than we could know according to the quantum's rules.

Instead we must write the state qwiff as a superposition of these two possibilities, UD and DU.[†] Now this superposition contains neither the state UU nor DD. In other words, the probabilities of finding the two electrons with either both spins up or both spins down has been reduced to zero. Now not all possible states are equally likely. Instead we find both UD and DU with equal probabilities of one-half each, but states UU and DD each have zero probability. The information in this correlated qwiff has been reduced. Calling this information $I(1, 2)$ and recognizing that there are only two possibilities, leaves

$$I(1, 2) = \log(2) = 1$$

Now we can compare the information of the correlated qwiff, $I(1, 2)$, with the sum of the informations of the uncorrelated qwiffs of the separate electrons, $I(1) + I(2)$. Since $I(1) + I(2) = 2$, the result is

$$I(1, 2) < I(1) + I(2).$$

(The $<$ means "is less than.")

The result, that correlated information or cooperative information is always less than the sum of uncorrelated informations, is a general result. The higher the correlation, the more particles involved in the cooperation, the lower is the information in comparison to the sum of the separated uncorrelated qwiffs representing the particles.

And it is here that brains and minds overlap. Here entropy or information, I believe, takes on meaning. I offer the hypothesis that meaning is correlation. The greater the correlation, the more significant the meaning, i.e., the greater is the sense of truth to be found. For example, can you think of any thought that has uncompromising meaning? By this I mean a thought that means what it says universally, with little or no disagreement. If such a thought exists it clearly must correlate a lot of things! It must be something that is universally acceptable because it applies to many, many diverse and wide-ranging concepts, ideas, objects, etc.

One such thought is "number." The concept of "number" is clearly universal. Few would disagree with the meaning of "two" or "three." "Number" must be a concept that correlates with everything possible in the universe. That is why it is such a powerful concept.

Another thought is "love." Here, although everyone knows what love is, few can define it numerically or in any other way. Actually the same is true of "number," but we have grown used to manipulating numbers while having seemingly little control of love. "Hunger," "thirst," "death," etc., are powerful words rich in meaning. When we correlate human experiences with each other, we experience a "sense of meaning." This sense is much like our other senses.

---

†And because of another rule contained in the quantum world game, we must not add the two possibilities but subtract one from the other. In the next chapter we will look at the reason for the minus sign. In what follows here it plays no role.

It differs from them in that it is an internal sense, one that seems to go on inside of us. Some may refer to this sense as intuition, or perhaps intelligence, or insight, a word I like because it resembles the sense of sight.

This internal experience arises from the correlation of qwiffs, the qwiffs representing microsystems in our brains. These microsystems, as I pointed out earlier, could be molecular groups attached to enzymes governing the firings of neurons. Or they could possibly be found on the even smaller scale of atoms, even possibly electrons themselves. I believe that it is the electrons within the closed orbital shells of atoms, probably hydrogen atoms contained in our water molecules, or in carbon atoms, that are the fundamental microsystems used to correlate information and thereby reduce it.

Indeed, the spin of the electron makes an ideal nondissipative *Turing machine.* (A Turing machine is a fundamental computer. It is capable of assuming one of a finite number of states and following a set procedure to alter those states, usually by reading and writing on a memory record.) Turing machines, named after the brilliant mathematician and computer scientist Alan Turing, are slow but powerful computers. Any computer built today and operating digitally can be replaced or simulated by a Turing machine.

Since electrons are two-state "devices" (spin U or spin D), they are binary elements, ideal for digital computer operations. Could it be that electrons on inner shells of carbon atoms are our "memory and microprocessor units"? Our PROMS, RAMS, and microprocessors?

Each positive thought might be positive and healing, loving and fulfilling, simply because it correlates electrons in the "carbon units" contained within our brains and nervous systems. On the other hand, destructive thoughts, or explosive behavior, "break out" of the bondage of correlations, freeing the electrons from their correlative, and therefore restrictive, bonds. Information is released in this. The entropy of the universe, like the gloating devil waiting for his due in a Faustian epic, is satisfied.

PART THREE

PSYCHOPHYSICS:
THE MIND OF MATTER

The vision that is coming, the new "Western psyche," projects a powerful and correlative experience. The new order brings matter and mind together. In this part I will explore the "new psychology" and propose axioms for meaning, the quantum laws of consciousness and mental states.

6

Love and Hate, Light and Matter

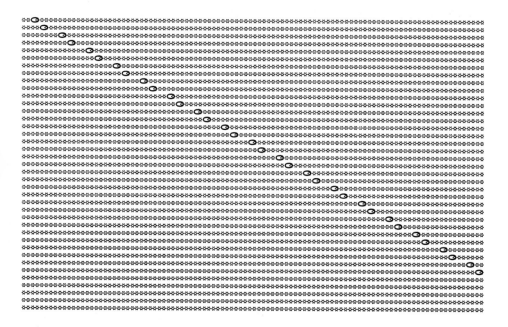

Much madness is divinest sense
to a discerning eye;
Much sense the starkest madness.
'Tis the majority
in this, as all, prevails.
Assent, and you are sane;
Demur, — you're straightaway dangerous,
And handled with a chain.

Emily Dickinson

Brain/Mind in the "New Physics"

Our brains follow the same basic laws of the whole universe. These laws are quantum mechanical. In our brains dance electrons and photons — the particles of light responsible for communication between electrons. The dance of electrons with light makes up our minds and provides the stability of our 85-percent water-filled brains. Thus it is that mind will not be found in any physical pattern of our brain material, but instead in quantum wave functions. In this chapter the delicate and powerful choreography of this dance will concern us. To understand the dance it is necessary to look at the statistics of electrons and photons. These statisics are profoundly "psychic" and as nonmechanical, non-Maxwellian (remember the bell-shaped Maxwell curve of classical statistics) as any statistics could be. They even bring to light the most compelling sense of reality we have and show that even that sense is an illusion. That illusion is the experience of things being separate and possessing established identities. Following on Freud's ideas that our emotional states are based on the unconscious battle between the life and death instincts, and that "as a result of the combination of unicellular organisms into a multicellular form of life, the death instinct of the single cell can successfully be neutralized and the destructive impulses be diverted on to the external world," it is possible to form a quantum-physical model of love and hate, the "game" of the ego.

By combining into a multicellular life form, single cells are able to form correlated qwiff patterns, which as we saw in Chapter 5, reduce entropy and information, producing more physical order. There is also another Freudian insight here. Freud described the ego as "the surface of the mental apparatus constituting the unconscious." I suggest that each atom behaves as if it had a tiny ego of its own. When atoms combine to form molecules, they lose their individual identities as atoms and take on the group identity of the molecule, thereby increasing their mutual surface area. However, when two atoms, each with a volume V combine, the combined volume becomes $2 \times V$, but the mutual surface area does not double. It increases by only a factor of 1.58. If three atoms combine, creating a mutual volume of $3 \times V$, the mutual area is only doubled.

134

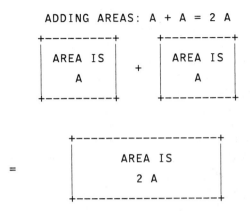

ADDING AREAS: A + A = 2 A

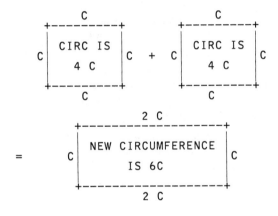

BUT ADDING CIRCUMFERENCES: 4C + 4C > 6C

Figure 27

Adding areas always produces a reduction in the combined circumference. Similarly, adding volumes always produces a reduction in the combined surface area surrounding the whole volume. This illustrates the ideas that combining or correlating always reduces information and entropy and that a group ego is always "smaller" than the sum of its individual egos.

This reduction in surface as a result of adding volumes is a geometric fact. The molecular ego, if attributed to the surface of the molecule formed by the individual atoms, is greater than the ego of any individual atom but less than the sum of all the atomic egos forming the molecule. This is strikingly similar

135

to the correlated information or entropy of the molecule compared to the sum of the individual entropies. (A similar analogy has been drawn by J. Beckenstein in his discussion of the entropy of black holes — gravitationally crushed stars formed when the self-gravity of a star exceeds all possible pressures sustaining the star from collapse. Beckenstein and Hawking discovered that the surface area of two black holes was less than the sum of their separate areas. This means that when black holes combine, they liberate entropy.)

With each increase in the number of members forming a group as, for example, employees of a corporation, the group ego grows but is always less than the sum of separated egos. As a member of such a collective, the individuals lose identity and a sense of individual responsibility. The group performs in a manner that no individual is capable of. Because the group ego is less than the sum of its parts, the group does not "remember" its parts, its individual behavior patterns; it cannot "vibrate sympathetically" with its individual members. Just as the IRS can and does act in a manner that no taxpayer would be allowed, and does so without any "human sense of right or wrong," the group created by its individuals performs in a new and extraordinary manner compared with the behavioral patterns of its members.

Societies, governments, powerful corporations have a similar group identity and ego structure. During wartime individuals commit murder by killing each other without feeling any apparent individual guilt. The most notorious example of this was the Nazi war machine. Each "war criminal" swore to his or her own innocence, saying, "I was only following orders."

The human individual made up from living cells also exhibits an ego structure greater than any cell but less than the sum of the egos of those cells. Ego and information are formed from the same stuff. The reduction of the cell's death instinct by the formation of the thinking brain that senses its own death as its own entity — which each of us articulates when we say, "If I die before I wake . . ." — is, I believe, related to the surface area of the organs formed by the cells. And cells are in turn formed from atoms, which get their identities, their egos, from electrons that learn that they are not alone by "dancing" with light.

The Physics of Hate and Fear

The latest experiments at SLAC (Stanford Linear Accelerator) confirm it. Einstein's $E = mc^2$ asserts it (E is energy, m is mass, and c is the speed of light). It is no exaggeration to say that matter is "trapped light." The processes of "light trapping" require a lot of light energy, however. That c in Einstein's equation is squared, multiplied by itself, and it is a big number on its own. That means a lot of E (light energy) goes into making a small amount of m (mass). In the study of quantum electrodynamics the most fundamental process is the interaction of the tiniest particle of matter known to humankind, the electron, with the photon, a particle of light.

136

In a similar manner I believe that there exist fundamental interacting "units" of the psyche. I call these units *morphemes.* Each morpheme interacts through the creation of a quantum correlation. Morphemes are the correlation of the psyche just as electrons are the light traps of physical space. Each morpheme thus behaves like an electron. However, these morphemes do not move through space and time but propagate through sensation and thought. By making vertex maps of the interactions between morphemes, the basic "feelings" of love and hate can be understood. All other feelings can then be understood in terms of these basic feelings. All feelings arise from particular combinations of the basic emotions of love and hate. And these feelings are fundamental to the electron's behavior in physical space manifesting as quantum statistics. Electron self-hatred gives rise to atomic structure. The fundamental electron-photon interaction is called a *vertex* and appears as shown in Figure 28.

THE VERTEX: THE UNIVERSE'S FUNDAMENTAL PROCESS

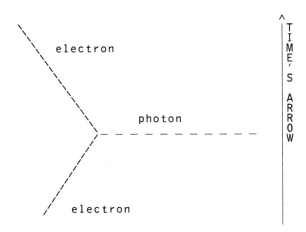

Figure 28

A fundamental vertex shows a single electron moving up the page in the direction of time's arrow. The electron has been diverted from its movement to the right by the emission or absorption of a photon—a particle of light. The electron then moves off toward the left of the page. This vertex is the most fundamental process of quantum electrodynamics—the interaction of matter with light. This never occurs by itself, but only in pairs: Two vertices do an interaction make.

By combining vertices, physicists put together a space-time map of the history of electrons interacting with each other. Each vertex is either the emission of a photon or the absorption of one. With each emission the electron loses part of its mass. With each absorption it gains in mass.

Vertices pointing to the left or right represent "ordinary" processes where the electron is merely disrupted from its normal journey through space and in time. Vertices oriented up or down, however, show a different thing (see Figures 29 and 30).

PAIR ANNIHILATION: WHEN MATTER TURNS TO LIGHT

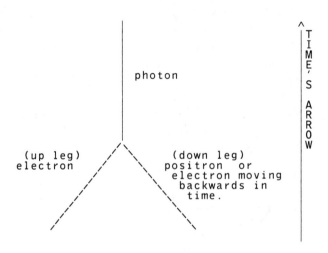

Figure 29
An up vertex shows a single electron moving up the page in the direction of time's arrow. The electron has been diverted from its movement forward in time and to the right by the emission of a photon—a particle of light. The electron then moves off backward in time and continues toward the right of the page. This vertex is the same process of quantum electrodynamics—the interaction of matter with light as shown in Figure 28—only turned around by one quarter turn. To an observer flowing forward in the direction of time's arrow, the electron appears in both places right and left of the oncoming vertex event. The right or down leg appears as an up leg for a positron, a particle of antimatter.

An up vertex means that the electron is "turned around in time" and vanishes from the time stream. The down leg of the vertex is the electron traveling backward in time. This appears to an orderly viewer experiencing one-way time as the annihilation of an electron with a particle of antimatter, the *positron*.

A down vertex shows the opposite process from an up one. Here an electron moves backward in time along the left down leg and turns around,

PAIR CREATION: WHEN LIGHT GETS TRAPPED INTO BECOMING MATTER

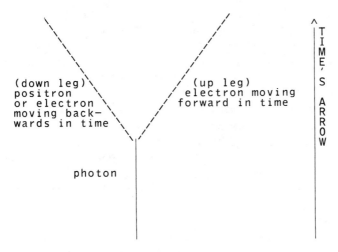

Figure 30

A down vertex shows a single electron moving down the page in the opposite direction of time's arrow. The electron has been diverted from its movement backward in time and to the right by the absorption of a photon—a particle of light. The electron then moves off forward in time and continues toward the right of the page. This vertex is the same process of quantum electrodynamics—the interaction of matter with light as shown in Figure 28 — only turned around by one quarter turn. To an observer flowing forward in the direction of time's arrow, the electron appears in both places right and left of the earlier vertex event. The left or down leg appears as an up leg for a positron, a particle of antimatter.

correcting itself as it goes forward in time up the right leg of the vertex. To a normal viewer this process is one of creation of matter and antimatter, a so-called pair creation.

These processes occur at each instance at every place in the whole universe. The emptiest corner of space, when viewed with the tiniest timepiece of time, shows that space-time dances in a constant and ever-changing cosmic light show, with electrons and positrons appearing and disappearing with awesome regularity.

Near the nuclei of every atom, the processes are even more intense. Somehow at the beginning of the universe an excess of matter was produced that has been trying to get back to its origins as pure light ever since the big bang. Matter, electrons, seeks to return to the blessed, timeless, and spaceless state of pure light.

Perhaps a note of explanation is due here. The fact that each electron possesses an identical electrical charge is, I believe, simultaneously a cry for annihilation and an assertion of identity. This annihilation occurs when an electron recombines with a positron — the antimatter partner in the process of matter creation. The positron contains a positive charge equal in magnitude but opposite in sign to the electron's negative charge. Opposite charges attract each other. Thus the electron is at once attracted to its antimatter partner and, as in the proverbial story of the femme fatale, that attraction repels it and defines its existence as a particle — a speck of matter.

Our brains, loaded with "real" matter, are storage arenas for this cosmic light show. Our electrical activity, which we call life, is rooted in the "death wish" of all electrons: to return to the state of oneness with the universe of light. All electrons want to be the same thing — to be light again.

Thus the electron's very existence is defined by its desire and its fear to return to the timeless, spaceless state of pure light. Paradoxically, this "inherent feeling" is its energy, its mc^2, and thus constitutes its individuality. Without this desire, or electrical charge, to be one again, to be alone — all one, each electron would lack identity — the ID card that says I am one.

The electron's electrical charge is a cry for the return to the void. It hopes and fears to attract its opposite, its antimatter partner, the positron, in the dance of death — the return to light. Electrical charges attract if they are opposites and repel if they are alike. Thus two electrons brought together will "feel" each other's presence as a pushing away from each other. It is only through the presence of a positive charge (opposite to the negative electron charge) that this battle between two electrons can be neutralized.

Such is the case in the balance between electrons and a central nucleus in an atom. The positively charged nucleus attracts electrons as a hive of honey draws bees. Carefully avoiding each other, the electrons "buzz" around the powerfully charged nucleus of the atom. The nucleus contains as many positive electrical charges as there are electrons "buzzing" around it.

But this isn't the whole story. If it were, life would be totally impossible. All chemical activity would have ceased unless the temperature of the bath surrounding that activity was millions of degrees. The reason for this is twofold. First, the atom follows the rules of quantum physics. This means that the electrons in the atom must exist in *quantum states*. These states are determined by the boundary conditions existing within the atom. Going back to Chapter 1 for a moment, remember the example of two particles in a tube? The walls of the tube were the boundary conditions limiting the behavior of the qwiff in the tube. That qwiff acted as the probability wave for locating each particle. Wherever the qwiff was most intense, the probability of finding both particles was greatest. Each particle was acting as a whole, sensing the presence of the other as a correlation. This pattern is called a quantum state.

Now, in such a state the uncertainty principle holds rein. Whatever is knowable appears as a quantum number in a quantum state. In an atom there

140

are four numbers associated with each quantum state. These numbers are typically labeled:

n— the energy quantum number, also called the shell number
l— the angular momentum quantum number, also called the orbital quantum number
m— a component of the angular momentum usually thought of as the projection, or "shadow," of l along a specific direction in space
s— a component of spin — the intrinsic angular momentum of the electron whose state is under question

Together these four quantum numbers specify all that can be specified about each electron in an atom. Taken all together we have four "bits" of knowledge about each electron, n, l, m, and s. Now we come to the second reason for life's possibility and the presence of all chemical activity. Put very briefly, it is due to electronic "self-hatred." Without this hate the four "bits" of life and chemistry would indulge themselves by keeping n as small as possible while forcing l and m to be as large as they can be. This would make breathing and eating, indeed any chemical activity necessary for life's processes, impossible.

In 1925 Wolfgang Pauli proposed the first concept of electron self-hatred. He called it the *principle of exclusion.* It says quite simply that no two electrons in the universe can have identical quantum numbers. Each and every electron is doomed to solitude whenever it exists. Another way to state this is: No electron may enter into a state already occupied by another electron. It is Pauli's exclusion principle that keeps atoms from getting too fat and electrons from falling into the lowest energy state, $n = 1$. Also, each and every property of the elements owes its uniqueness and its similarity to the Pauli exclusion principle.

Each electron behaves like Greta Garbo, crying to all the other electrons surrounding it, "I vant to be alone."[†]

The electron's loneliness and "her" insistence on "doing her own thing" and maintaining her separate identity by having unique quantum numbers is, I believe, the origin of our own egos, self-hatred, and, when reflected onto the outside world, our tendency toward destruction. It is the origin of the fear mechanism, that uncomfortable feeling that always crops up when we are put into strange surroundings, especially with strange people around us.

Does the electron really "cause" all human fear? In the same sense that a group ego, i.e., a corporate ego, is always less than the sum of its individual egos but greater than any individual's ego, a group fear is composed of the fears of its constituents. Thus any individual person, animal, or plant, being made of electrons, contains electrons' fears. These fears form the composite fear experienced by that individual as the individual's fear. The fear of death, which ultimately is the base fear of all other fears, thus originates in the electron's

[†]Her most famous line, which she uttered in the classic film *Grand Hotel.*

fear/identity — its annihilation dance with death.

Our death is linked to the quantum electrodynamical process of pair annihilation (see Figure 29). Since all vertex processes (see Figures 28, 29, and 30) are really the same process, annihilation, creation, and interaction are the origins of our life and our death.

Our brains are "loaded" with electrons. These impossibly tiny point-size particle-waves are responsible for all brain activity. Without these little despicable characters running around inside of our heads, all yelling for individual attention and getting it, two precious neural atoms, sodium and potassium, would be nearly impossible to ionize. These two atoms must become ions; each must lose one electron to surrounding chlorine and other atomic electron acceptors (who then become negatively charged ions). These two ions make up a team used by our neurons in a "chemical pump" to propagate electrical impulses along a nerve cell. By exchanging positions across the neuron's membrane, an electrical pulse of energy can be sent along the axon of the neuron. Since both potassium and sodium have low ionization thresholds, it is relatively easy to produce them in our warm bodies. However, potassium and sodium, although chemically similar, are quite different in atomic mass, potassium the heavier one by a factor of nearly two. Thus potassium moves across the membrane barrier sluggishly while sodium, having a smaller mass, moves more freely. This difference in mobility probably accounts for nerve-pulse propagation. Nature has created a perfect mechanism for electrical communication in our bodies by having potassium and sodium possess nearly equal electronic but widely different inertial behavior. Without the electron's exclusivity, potassium and sodium would not provide the mechanism of nerve conduction.

It is this "peer pressure" from electrons that totally accounts for the electronic and chemical behavior of all of the known elements in the physical world. Without this exclusivity — this, as I call it, self-hatred or need for aloneness of electrons — all atoms would more or less appear chemically and elecronically inert, doomed to chemical isolation and, paradoxically, loneliness.

And there is more. The peculiar self-hatred can generate a "psychic" or telepathic communication or correlation between electrons. This correlation arises from two properties posessed by each electron: electric charge and spin.

Through electrical charge each electron "knows" about the other electrons: The charge generates a spherically symmetrical electrical field surrounding it that provides for the interaction between electrons. In the vertex picture mentioned earlier (page 137), when looked at in fine detail, this field appears as the photon line extending from a single vertex.

But this field is not the psychic awareness referred to here, but a necessary precursor correlating the electrons. The psychic awareness arises in what is called the electronic *spin-space.*

It is possible to show a spin-space diagram for an electron similar in appearance to Figure 13, only this time there are only two possibilities in the "box." The spin-space diagram for an electron in the spin-up state appears in Figure 31 or, if shown in a vertical column — as in Figure 13, in Figure 32.

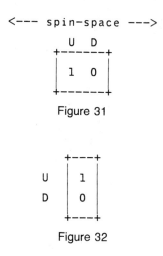

Figure 31

Figure 32

With two electrons that are uncorrelated in spin-space we can show a diagram similar to Figure 14 by making the spin-space for one electron vertical and the spin-space for the other horizontal, as we did for the two particles in the tube described in Chapter 1. The diagram appears in Figure 33A.

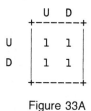

Figure 33A

In the above we have all possibilities equally likely. We could find with equal chance both particles with spin UP or both with spin DOWN or one with spin UP and the other with spin DOWN. This is quite similar to the situation found when two coins are flipped. There are four possible and equally likely situations of discovering the positions of the coin faces.

With correlation, however, the randomness shown above is reduced. The requirement of the Pauli exclusion principle makes it impossible to have both electrons in the same spin-space at the same time. This impossibility introduces

143

a change in phase in the qwiff, which appears as an antisymmetry in the diagram.

Since the electrons "hate" each other and since there is no way to tell one from the other, the two-electron qwiff must contain the two possibilities in such a way that the qwiff itself is antisymmetric. Antisymmetry was discussed earlier in Chapter 1 and shown particularly in Figure 17. This shows a spatially distributed antisymmetry that accounts for the impossibility of finding both particles on the same side of the tube or finding both particles with the same energy. Similarly here the antisymmetry appears clearly in the distribution of plus and minus signs and in the vanishing of the qwiff along the diagonal of Figure 33B.

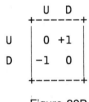

Figure 33B

Thus our two electrons must take on oppositely directed spins, one pointing up and the other down. But which has which?

Since there is no observable way to tell the difference between the two electrons, both possibilities must be present with equal probabilities. This leads us to consider the qwiff for the two electrons, and therein lies the psychic "rub."

Through the earlier Pauli exclusion principle, or electronic self-hatred, there arises simultaneously both the "primal ego" and the origin of fear, both originating in the pointlike structures of matter of which we are made. The electron fears its own annihilation, which will result if it interacts with its antimatter "shadow" particle called the positron (which you recall is designed exactly like the electron, with the tiny exception of having an electrical charge opposite but equal in magnitude). Of course, the result of that annihilation is the return to light. The space in which the electron finds itself, particularly when it is in interaction with other electrons, is continually bubbling out and reabsorbing virtual electron-positron pairs, which gives space a dynamic annihilation-creation character that bathes all matter, a constant remembrance that all matter is made of trapped light.

In a similar manner, annihilation and creation processes occur in the psyche. For every "positive" morpheme created there will appear a "negative" morpheme, or "antimorpheme." With each feeling of love arising there is therefore a feeling of fear. No one loves without remembrance of the fear of the annihilation of the ego.

The Pauli exclusion principle may result out of the annihilation of the fear "instinct" possessed by the electronlike particles (called *fermions* after Enrico Fermi and Paul Dirac, who first discovered that their exclusion property resulted in new statistical configurations). Thus fear *is* matter and matter is ultimately free as light. All human fears and animal fear mechanisms — indeed all living fear in all plant and animal life — are similarly traced to electronic self-annihilations. The only escape is in remembrance of our true source. We are all beings of light from the lowliest to the highest among us, from the slugs to the astronauts.

The Physics of Love

How does "love conquer all"? Remarkably it is the other half of the vertex interaction, the photon, that is responsible. To understand this it is necessary to look at photon statistics — the "physics of love."

There is an old puzzle about cards and a hat that goes something like this: You have a hat containing three cards. One card is red on both sides. The other is blue on both sides and the third is red on one side and blue on the other. You shake up the hat and hold it in such a manner that you cannot see inside. You pull one card from the hat, carefully exposing one side so that you cannot see the other side. The exposed side is blue. What is the probability that the other side is also blue?

You can go at this little puzzle in several ways. You might reason it out along the following line: Clearly there are three cards in the hat. Only one of the three is blue on both sides. Therefore I have only one chance in three of getting that card. The answer is 1/3.

Or you might go at it this way: Since there are three cards in the hat and since I see that I have a card with a blue side showing, I clearly have not chosen the red-red card. I either have the red-blue card or the blue-blue card. That means a fifty – fifty split in the odds. So the answer is 1/2.

Surprisingly both of the above answers are incorrect. The correct answer is 2/3. In other words, if you see that you have a blue side showing the odds are two out of three, or two to one, that you have chosen the blue-blue card.

Why? Because you know more than you think you know, or, perhaps put better, there is a greater chance of both sides being blue statistically speaking if you know that one of the sides is already blue.

Before I show you how this works I want to point out that this little rebus has much to do with what I call the "psychic" connection between photons and is the primary statistical reason that a laser works the way it does. It also will explain a predominant feature of photons — particles of light. In contrast to the annihilation-fear behavior of electrons, they all desire to be in the same quantum state. The reason for this desire to "condense" into one state is that photons with identical energies, frequencies, momenta, and spin tend to produce a symmetric state when they are correlated. This indistinguishability of photons alters the crapshoot of God's opening move in the game of the universe.[†]

The reason that the probability of the card being blue-blue is 2/3 follows along these lines: First of all, there are actually six possible faces of the three cards, three red and three blue. Let us label them according to Figure 34.

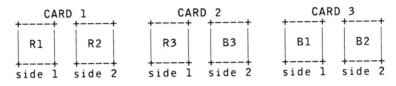

Figure 34

Thus when you first pull out a card and see a blue face, you are actually looking at B1, B2, or B3. This gives you three possible experiences depending on which blue face you see. B1 yields the other side as B2. B2 yields the other side as B1. B3 yields the other side as R3, a red face. Three possibilities only, and in only one of these cases is there a red hidden side. Therefore the odds are two to one that if you have a blue face facing you, the hidden face is also blue.

As I said, you actually have more information when you have a blue face exposed than you think you have using "common sense."

To see the connection of this puzzle with a photon and its "identity," or "lack of ego," crisis consider a laser tube containing only three atoms, each in an excited state ready to emit a photon of light, hypothetically shown in Figure 35.

LASER TUBE WITH THREE EXCITED ATOMS

Figure 35

The atoms A1, A2, and A3 are all excited (that's what the * symbolizes). To make life simple, suppose that each atom is capable of emitting its photon only toward the right end of the tube, and in only one of two possible "polariza-

†"Let There Be Light, and there was Light."

146

tion," or spin, states, which I will label as "red" or "blue" (to make the analogy with the card puzzle more understandable). This means that each atom left alone can emit either a red or a blue photon. The odds are therefore fifty – fifty that an atom, totally alone, will emit a blue photon.

Suppose that A1* emits a blue photon. We write this as

$$A1^* \longrightarrow A1 + B \text{ (blue photon)}$$

That is, it spontaneously emits and, with 50 percent probability, sends a blue photon on its way toward the half-silvered mirror end of the laser tube. What is the probability that atom A2* will also emit a blue photon? Let's look at Figure 36.

LASER TUBE WITH TWO EXCITED ATOMS AND ONE BLUE PHOTON

Figure 36

The answer would appear to be 50 percent. After all, didn't I just say that each atom can emit either a blue or a red photon with equal probability? Well it turns out that the answer is actually 2/3, and for the same reason as in the cards-in-the-hat trick. There are not just photons present in the laser tube; there is also knowledge, more knowledge than you think would be there at first glance.

To see this we again must consider the possibilities. There appear to be four distinct cases, as shown in the Figure 37.

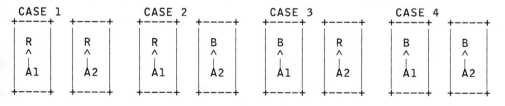

Figure 37

Each case shows an atom A1 or A2 emitting a photon R or B. Of course, we can dismiss cases 1 and 2 because we know that

$$A1^* \longrightarrow A1 + B$$

That is, the atom A1 emitted a blue photon. That still leaves cases 3 and 4. Since there are only two cases, this means that the probability of A2 emitting a blue photon should be 50 percent, completely oblivious of what atom A1 has done.

But now the ego, or identity, crisis enters the scene. As if by magic the universe asks us to reconsider what we "think" we know. Do we really know that it was A1 that made the blue photon? If the answer is that we had no way of telling which atom had the blue emission, that it was perfectly possible that either atom A1 or A2 emitted blue, then the probabilities change and become those of the cards-in-the-hat trick. Consequently, the atoms in this case do exhibit a surprising "psychic" link, which forms the basis of a remarkable synchronicity.

So, given that we really can't determine which atom emitted the blue photon, only that one of them did — i.e., that for sure there is at least one blue photon present — the number of possible cases present is altered.

If we restate the problem as, "Given that there are two photons present, emitted by two atoms (one photon emitted from each atom without our knowing which atom emitted which photon), how many possible cases are there?", we find that there are only three possible cases, as illustrated in Figure 38.

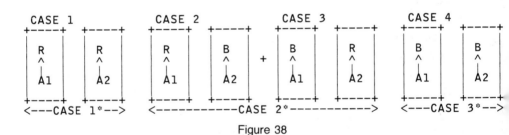

Figure 38

Cases 2 and 3 coalesce into case 2* while cases 1 and 4 remain the same (just relabeled). This coalescing is a quantum-physical phenomenon that must occur whenever two possibilities cannot be distinguished. Since we cannot tell which atom emitted which photon, we need not distinguish A1 from A2. The above then becomes Figure 39.

And we can see that in cases 2 and 3 we are actually counting the same physical situation twice. (Case 2 is the same as case 3. To see this, compare Figure 38 with Figure 34, and by analogy think of case 1* as the same as card 1, case 2* as the same as card 2, and case 3* as the same as card 3. Case 3 is actually a duplicate "card" of case 2.)

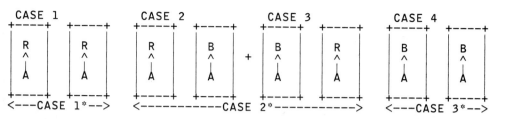

Figure 39

PHOTON POSSIBILITIES

CASE 1   CASE 2   CASE 3

```
+---+  +---+      +---+  +---+      +---+  +---+
| R |  | R |      | R |  | B |      | B |  | B |
+---+  +---+      +---+  +---+      +---+  +---+
```

Figure 40

Thus with two indistinguishable photons present, there are only three possible situations, as shown in Figure 40.

Given that one photon is blue is analogously the same situation as pulling a card from the hat and seeing a blue face. Since there are only three possible ways to realize a blue photon (case 1 doesn't count) and in only one of those ways is the other photon red, the remarkable fact is that with one photon known to be blue, the second photon is suddenly limited in its possibilities: It has only one chance in three of being red.

Now, since there is no causal physical connection between the photons (in fact they could both be emitted simultaneously), there is no possible mechanical explanation of this phenomenon. The only "thing" causing this remarkable "adjustment" in the odds is our knowledge (or, better, our lack of knowledge) concerning which atom emitted the known blue photon. Here mind enters the physical domain with a profound effect.

With the inclusion of a third excited atom ready to emit and the knowledge that there are actually two blue photons present, the third photon is limited even further (as shown in Figure 41 on page 150).

We can visualize the possibilities with triangles showing four possible cases (Figure 42 on page 150).

Cases 1 and 2 need not be considered (since the bb combination cannot occur in these cases), so we either have case 3 or case 4. We see that there are four "sides" in which the combination bb (standing for blue-blue) occurs in

149

LASER TUBE WITH ONE EXCITED ATOM AND TWO BLUE PHOTONS

Figure 41

PHOTON POSSIBILITIES

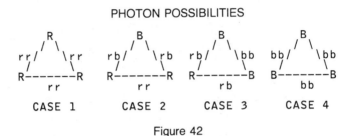

Figure 42

cases 3 and 4. The other two "sides" are rb.

Thus, referring back to the "hat" trick, if we were to place these four triangles in a hat and pull out one triangle, noticing only that the side facing us was a bb side (i.e., that the two corners facing us were both blue), what is the probability that the third corner is also blue? In three out of the four side examinations in which bb are seen, the hidden corner is blue. In one out of the four side examinations in which bb are seen, the hidden corner is red. Thus we conclude that with two blue photons present, the probability is 75 percent that the third photon is also blue.

In other words, if you were to pull one of the two triangles from the hat where a bb could be seen, the chances are that three out of four times you would be holding case 4 in your hand. Similarly, when there are two blue photons present in the laser, the odds favor the third photon being "true" blue, too.

With four photons present and three of them blue, the odds are four out of five that the fourth is also blue. With fifty photons, the odds are fifty out of fifty-one. In an actual laser with billions upon billions of "blue" photons continuously emitted and new, excited atoms being created from an electrical "pumping" source, the probability of "blue" emission is virtually certainty.

This "psychic" sense exists between identical particles whether they are electrons or photons. For electrons it is manifested as "self," "ego," "hatred," desire to be alone, and is the root cause of the Pauli exclusion principle described earlier. Indeed it appears that there exists a whole class of particles that exhibit exclusion. These include electrons, neutrons, protons, all nuclei with an odd number of nuclear constituents, quarks, and neutrinos. Thus, for example, all protons exclude each other. These particles are called fermions and are said to obey Fermi-Dirac statistics, as I mentioned earlier.

There is also a class of photonlike particles. These include all nuclei with an even number of nuclear constituents, photons, electron pairs, pions, and all particles composed of an even number of constituents that are so tightly bound to each other that the composite can be regarded as a "particle" in all inter-actions. These photonlike particles are called *bosons* and are said to obey "Einstein-Bose" statistics. (The names of Einstein and Bose are used because these physicists were the first to discover the boson's statistical properties.)

Now there is one primary element determining whether a particle is a boson or a fermion and that is the form or symmetry of the qwiff describing it when it is in interaction with another identical particle. For two identical fermions, the qwiff exhibits an antisymmetrical character as illustrated in Figure 43.

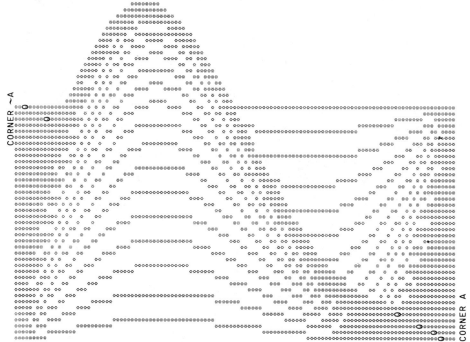

Figure 43
The physics of hate for two particles: an antisymmetric star wave produced by
Fermion-Pauli exclusion.

Now this qwiff does not occur in ordinary space. Instead we are showing it in Hilbert space (similar to the figures shown in Chapter 1 for two particles in a tube) — a space of our minds. Figure 43 is a three-dimensional representation (as shown in two-dimensional contour style with numbers marking the intensity, or "height," of the qwiff in Figures 14 – 21) of the qwiff for two identical particles in a tube shown in Figure 17. Each * marks the probability-wave amplitude that particle one is located at the position in the tube measured from the top of the page down to the * and particle two is simultaneously located at the position measured from the side of the page across to the *. The "cobblestones" (shown as ovals or sidewise zeros) mark the position where both particles have the same location in the tube.

Here there is a *solid-core* interaction between the particles. The imaginary line along the cobblestones from corner ∼A to corner A represents both fermions in the same place at the same time. The solid-core interaction forbids both particles occupying the same location at the same time, so the qwiff is zero or flat at each cobblestone along the diagonal line. To the right and above the cobblestone line the qwiff dips below the plane into a deep valley, while to the left and below the line the qwiff rises to a high mountain. The valley is as low as the mountain is high.

For a pair of bosons the qwiff appears to be symmetrical, as shown in Figure 44. It is the antisymmetry that is responsible for the fermions' exclusivity and the symmetry that is responsible for the bosons' "condensation" or desire to all be in the same state together.

In Figure 43, the case of two fermions, the qwiff is always zero along the diagonal line drawn from corner A to corner ∼A. In Figure 44, the case of two bosons, the qwiff is never zero (except at one point in the very center) along this diagonal line, but instead takes on its maximum extreme values. Here we find that the highest peak and lowest valley occur along this line indicating that the probability is highest whenever both particles are simultaneously positioned about one-fourth or three-fourths the length of the tube. The cobblestone diagonal line marks the simultaneous occurrence of both bosons at the same place at the same time, something that tends to occur often for bosons.

The photons in the laser tube are bosons and so tend to "psychically condense" into the same state. This "boson condensation" is the physical manifestation of a universal and very human quality — the feeling of love.

In other words, light is the lover. Light is the state where photons can experience the total lack of individuation marked so exclusively by electrons. Photons "love" to be in the same state. Like atoms in a laser tube tend to emit photons in the same state, resulting in the powerful, coherent laser beam. Thus electron "hatred," which makes the whole atomic arrangement possible, producing the "shell" structure of atoms through Pauli exclusion, is "relieved" in the stimulated emission of "loving," psychically communicating, photons.

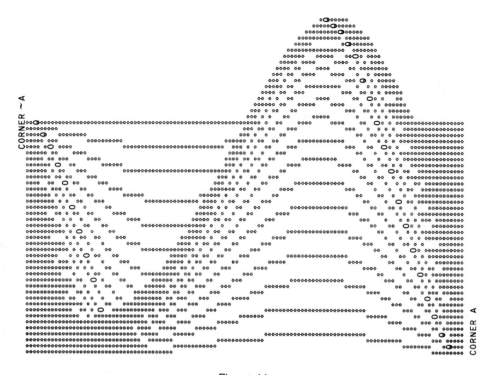

Figure 44
The physics of love for two particles: a symmetric star wave produced by
Boson condensation.

Our bodies, our sexuality, our entering into the state of loving communion is inherent in our photons and electrons. We are the "body electric." We are the "body of light." The dance of light and matter is the root "cause" of all emotional experience. It is also the root of thought. Thought and emotions go hand and hand as possibly complementary attributes of operational human experiences. We could say that through the universality of the quantum wave function all "things" feel and think, love and hate.

Love and Photons

The feelings we all develop for each other, the sense of ego, the desire for love and communion, that familiar feeling of falling in love, which tends to overwhelm us as teenagers, our fear of death, and even our spiritual sense of oneness with God could arise from the qwiff and its remarkable symmetry possibilities — symmetric or antisymmetric, boson or fermion. In this deep sense the human body-mind is that autonomically functioning aspect of spirit, or "qwiffness," which is ultimately the form and body of God. The qwiff and God are one. The aspects of qwiffness that arise as the universe of experience,

153

arise with the ego as the boundary line dividing the objective from the subjective, the symbol maker from the symbolic.

When two people fall in love, they are identifying with their inner psychic sense of oneness, with the symmetrical quality of the qwiff, the symmetry of light in Hilbert space, the condensation as one being of light. In the human brain learning occurs when there is identification with the process of synthesis — the new information is felt to be one and the same as the old information seen from a new light. Thus x is a number. An integral is exquisitely defined as "summing up." There is a process of surrendering going on.

Falling in love is only possible when each surrenders to the oneness of the impossible boson-condensing feeling of being in the same state at the same time. As wonderful as this is, in a similar manner a nation rises from the dust of defeat and with a cry for a "new order" enables each member of the nation to reidentify with the boson state — the third "rule" of such a state; and so comes "Hitler is Germany — Germany is Hitler."

Each "war criminal" swore that he or she was not responsible for any individual actions; each was simply an "order-follower." Each identified with "state-ego" not through the power of hate and fear, as many have come to believe who did not experience Nazi Germany, but through the power of love and supremacy. Hitler was Germany's greatest "lover."†

Our brains contain the perfect combination for the magnification of primal love-hate pointlike electron-photon dancing. Through nonlinear neuronal connections waves dance through the brain, allowing feelings and thoughts to arise if signals are reinforced from different locations within the brain. Wonderful, but what determines the pathways of reinforcement? Perhaps the brain operates holographically, using qwiffs as the waveforms.

The Brain Hologram

One of the most remarkable inventions made possible by the laser was the hologram. A hologram can be made only because of the coherence properties of light waves. And these coherences are the result of the photon statistics — the ability of two or more photons to enter into the same state. The more photons there are in a given state, the greater is the coherence of the light wave.

Our brains and bodies may also operate holographically. Recent experimental work described by Karl Pribram indicates that the processes occurring in the brain that we label as internal, such as feeling or hunger, are no different from those processes that register as our senses of the outside world. As Pribram puts it:

†Perhaps the difference between normal human love and the abnormal Nazi condensation lies in the normal love being made in the body of light and "Nazi-love" being made from pairs of correlated electrons.

154

Clinical neurological experience tells us that the localizing of a perceptual image is not a simple process. The paradoxical phenomenon of a phantom limb after amputation, for example, makes it unlikely that our experience of receptor stimulation "resides" where we are apt to localize it.

Even though it appears that we feel with our fingers and our toes, the evidence is overwhelming that the locations of those feelings are not taking place there. In a similar manner we see light that impinges on our retinas and hear sounds that disturb our eardrums, and yet we place the source of those sounds out in space and back in time to their appropriate space-time locations. We do not localize starlight at our retinas but thousands of light-years away. We do not localize the music of a concert pianist at the basilar membrane of the cochlea but at the keyboard of the concert piano.

We project experience outward from our brains and nervous systems. Since we have learned to do this since childhood for sight and sound senses, but not for touch and feelings, it is possible to learn to create a feeling in space where no "feelers" or skin even exist.

Pribram describes some important experimental work of physiologist G. von Békésy using touch. After describing some preliminary experiments using vibrators to simultaneously stimulate two fingertips, thereby generating the feeling of vibration between the fingers, Békésy, a Nobel laureate, wrote:

Even more dramatic than this experiment is the one in which two vibrators are placed on the thighs, one above each knee. . . . By training, a [subject] . . . can be made to perceive a sensation that moves continuously from one knee to the other. If the [subject] now spreads the knees apart he will again experience at first a jumping of the sensation from one knee to the other. In time, however, the [subject] will become convinced that the vibratory sensation can be localized in the free space between the knees.

Békésy writes about his experimental studies of how the mind creates external spatial locations of sensations.

I found the location of sensations in free space to be a very important feature of behavior. To study the matter further I wore two hearing aids that were properly damped so that the sounds could be picked up by means of two microphones on the chest and then transmitted to the two ears without change in pressure amplitude. Stereophonic hearing was well established, but a perception of the distance of sound sources was lost. I shall not forget my frustration in trying to cross the street during rush hour traffic while wearing this transmission system. Almost all the cars seemed to jump suddenly into consciousness, and I was unable to put them in order according to their immediacy.

How can we explain the localization phenomena? It could be that the ultimate explanation resides in the fact that the quantum wave function from which all experiences arise is internal and external at the same time. The world of experience is like a holographic world made from the interference patterns of qwiffs. Before we can grasp the significance of this we need to look at how a hologram works.

Although it is possible to simulate a hologram without using waves, wave interference is responsible for holographic reconstruction. Here is how a hologram works:

Suppose there exists floating in space a transparent object shaped somewhat like Figure 45.

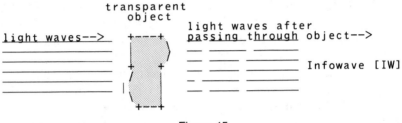

Figure 45

′ Lightwaves pass through the object and, as a result, are affected by its form or shape. Thus these waves are scattered and carry with them "piggyback" information about the size and three-dimensional shape of the transparent object. Now it is important to realize that these waves are coherent, consisting of photons "in love" with each other, and that, as a result, their ability to contain intelligent information is considerably enhanced.

This "infowave" has a precise mathematical form, which I shall designate as [IW]. The funny symbols [ ] remind us that this wave can be represented as a complex number.[†] (This is discussed more fully in Chapter 8.) If [IW] were to continue on its merry way without being recorded in some medium such as a photographic emulsion, no hologram could be produced. Certainly, if our eye were just to the right of the wave, we would see the shape of the object and "project" our image of it "out there" in space to the left, as shown in Figure 47.

[†]A number can be real, imaginary, or complex. A real number is any positive or negative ordinary number imagined to exist as a point on a line drawn across a page. An imaginary number does not exist anywhere on this "line of reals." Instead it is drawn as a point on a line perpendicular to the "line of reals." A complex number consists of a pair of numbers, one real and one imaginary. It appears as a point in the plane constructed from the cross of the "line of reals" with the "line of imaginaries." (See Figure 46.)

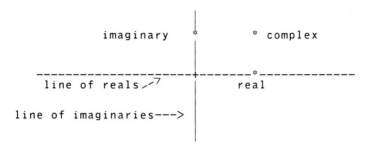

Figure 46

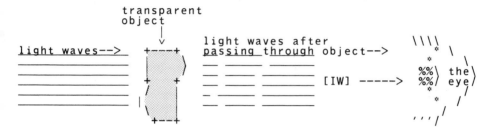

Figure 47

Our retina acts as a recording instrument for light. To do this it must record the light wave's energy. This is obtained mathematically from the [IW] by complex multiplication — the infowave complex number is multiplied by its "complex conjugate," written as [IW]*, a backward form of infowave itself. The result of the multiplication is [IW]*[IW], a self-referential product of the infowave with itself, in such a manner that a real and positive number is produced. This is important because the energy representing the recording must always be real and positive in order to have "reality." In fact any record is made in this manner. Ordinary photographic images are also a result of energy deposits made from [IW]*[IW].

In this recording of [IW]*[IW] something is actually lost, and that is the phase the light waves maintained after passing through the object. What we see is always produced from a recording and is therefore somewhat incomplete. To perceive [IW], the recording medium produces [IW]*[IW]. The act of registration is always irreversible and it simultaneously destroys information in order to record information.

But now suppose that part of the incoming lightwave is allowed to pass directly without going through the object (Figure 48).

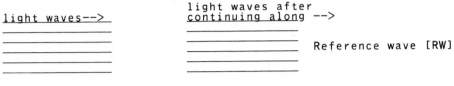

Figure 48

These waves constitute what is called a reference wave, [RW]. It, too, is a complex number containing a precise and simple phase relationship between its various waves.

When it is recorded on the retina it produces [RW]*[RW], which appears, because of the uniformity of the reference wave, as a uniform and overall darkening of the medium.

If the reference wave and the infowave are allowed to come together and add (something that waves can do, called *wave interference*), and if the superposition of these two waves,

$$[RW] + [IW]$$

is then recorded on a film emulsion, the resulting energy deposited on the film will be

$$([RW]* + [IW]*) \times ([RW] + [IW])$$

This complex multiplication results in four terms.

$$[RW]*[RW] + [RW]*[IW] + [IW]*[RW] + [IW]*[IW]$$
term 1   term 2   term 3   term 4

Now terms 1 and 4, because they are self-referential terms, have lost all phase information. Term 1, because it consists of the reference wave interfering with itself, will produce a uniform darkening of the emulsion. Term 4, because it consists of the information wave interfering with itself, will produce a small variation in this darkening, but will not reproduce any information (because that information is contained in the phase as well as amplitude variations of infowave). Thus only terms 2 and 3 are important because they contain the original infowave in relationship with the reference wave. Phase information is not lost here.

Phase information may be a new concept for the reader. To grasp its significance think of the reference wave as a sequence of grid lines or street signs and the information wave as falling rain. In order to know just how much

158

rain fell in any part of a city, one must determine how much fell in between particular grid lines or between certain streets and avenues. Multiplying any wave by its complex conjugate is like laying a negative photograph over its positive; the resulting image is washed out. But overlaying a negative grid over a positive rain-distribution photo brings out the rain-distribution information and shows it quite clearly. (See Figure 49.)

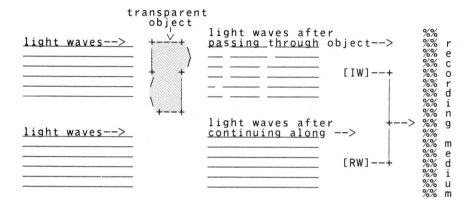

Figure 49

The medium records these four terms and thus contains a record of the whole three-dimensional transparent object, including the important phase relationships. In fact, it contains two such terms. Term 2 as recorded in the medium will, upon receipt of any light wave, reproduce a virtual image of the transparent object. Term 3 will reproduce a real image of that object.

A virtual image is the kind you see when you look at yourself in a mirror. The person in the mirror appears to be behind the glass. A real image is what you see when you watch a motion picture. The projector focuses the light passing through the recorded film image and recreates the image on a screen. Real images can always be focused on a screen while virtual images always appear to be coming from places where they do not exist.

When a new light wave impinges on the recording medium, it acts as a filter, allowing some light to pass through and pick up the imprint in the medium. However, unlike a photographic image, it is the interaction of the new light wave with the recorded interference pattern that is seen as a three-dimensional image. It is three-dimensional because the new light wave becomes modified by the wave-interference pattern recorded in the emulsion and appears thereby as the original light wave emitted by the object itself. (See Figure 50.)

159

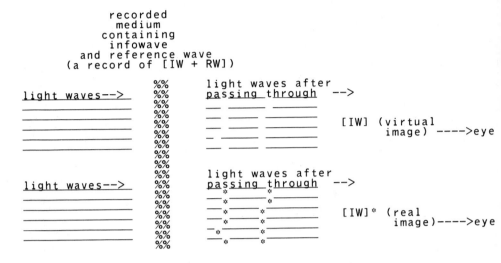

Figure 50

This light then will also consist of four terms each multiplied by the filtering factor of the medium. Again two terms will appear only as the modified light beam passing through. But the two other terms will contain the information originally recorded, multiplied by the intensity of the new light beam. These terms will be seen by the observing eye as a virtual image reconstructed in space and as a focused real image appearing either inside the witnessing eye or out in space just in front of it.

There are some other remarkable properties to all this. Each piece of the recorded medium, called the hologram, contains the interference pattern produced by adding the reference wave to the information wave. Each point on the object acts as a tiny sender of spherically distributed information waves, and these waves hit all points on the surface of the recording medium together with the reference wave. Thus each tiny jigsaw-puzzle piece of the recording medium contains information from all points on the object. This is where the idea of the part containing the whole arises.

Now if the medium is not "developed" so that it remains sensitive to further illumination, something further remarkable occurs. Suppose a second object is illuminated using the same reference wave. It will produce an information wave, [JW]. Again there will be energy deposited according to a similar formula with "JW" the information wave from the second object:

$$([RW]^* + [JW]^*) \times ([RW] + [JW])$$

This complex multiplication also results in four terms.

$$\underset{\text{term 1}}{[RW]^*[RW]} + \underset{\text{term 2}}{[RW]^*[JW]} + \underset{\text{term 3}}{[JW]^*[RW]} + \underset{\text{term 4}}{[JW]^*[JW]}$$

But now the medium contains two energy deposits that when added together result in

$$\underset{\text{term 1}}{[RW]^*[RW]} + \underset{\text{term 2}}{[RW]^*[IW]} + \underset{\text{term 3}}{[IW]^*[RW]} + \underset{\text{term 4}}{[IW]^*[IW]}$$

$$+ \underset{\text{term 5}}{[RW]^*[RW]} + \underset{\text{term 6}}{[RW]^*[JW]} + \underset{\text{term 7}}{[JW]^*[RW]} + \underset{\text{term 8}}{[JW]^*[JW]}$$

Again, following similar reasoning, terms 1, 5, 4, and 8 will, on reillumination (because they are self-referential terms), have lost all phase information: Terms 1 and 5, because they consist of the reference wave interfering with itself, will only appear as the modified light beam passing through. Terms 4 and 8, because they consist of information waves interfering with themselves, will produce a small variation in this intensity, but will not reproduce any information (because that information is contained in the phase as well as amplitude variations of the infowave). Thus only terms 2, 3, 6, and 7 are important because they contain the original infowaves in relationship to the reference wave. Phase information is not lost here.

The Algebra of Holograms

By regrouping terms 2, 3, 6, and 7 we see also something else remarkable occurring.

$$\underset{\text{term 2}}{[RW]^*[IW]} + \underset{\text{term 3}}{[IW]^*[RW]} + \underset{\text{term 6}}{[RW]^*[JW]} + \underset{\text{term 7}}{[JW]^*[RW]} =$$

$$\underset{\text{term 2}}{[RW]^*[IW]} + \underset{\text{term 6}}{[RW]^*[JW]} + \underset{\text{term 3}}{[IW]^*[RW]} + \underset{\text{term 7}}{[JW]^*[RW]} =$$

$$\underset{\text{term 2/term 6}}{[RW]^*([IW] + [JW])} + \underset{\text{term 3/term 7}}{([IW]^* + [JW]^*)[RW]}$$

Terms 2 and 6 add together and both multiply the complex-conjugated reference wave [RW]*, while terms 3 and 7 add together and both multiply the reference wave [RW]. This means that there is now contained within the recording medium an interference pattern between two waves IW and JW even though these waves were recorded at entirely different times. When a new light wave is passed through the "developed" medium after it has received both JW and IW, the eye will see the interference of two past events.

Holographic Remembering and Forgetting

Herein lies a mechanism for how the brain remembers and how it can forget. If the cerebral cortex is a recording medium capable of holographically recording qwiff waves, then associative memory may be the result of the interference of two or more information sources "witnessed" when an appropriate reference wave reverberates through the cortex.

In a similar manner, by using an appropriate "randomizing" second object, the image of the first object can be "forgotten" or wiped out. If, instead of using one RW, a series of RW's were used, a selectable variety of "memories" could be recorded for a rich, creative, associative thought pattern. Even a slight change in the direction of the reference beam alters the interference properties of the "memory record," enabling a "search-scan" to reproduce a series of memories. The temporary halting we all experience as we "scan" our memories looking for a forgotten name may indeed be such a scan of the three-dimensional holography we all carry in our heads. The sudden remembrance of a forgotten fact when we relax our minds and stop our fervent searching may be due to the qwiff beam randomly scanning the cortex record.

The inability of elderly people to memorize new information may be due to the inability of the cortex to elastically modify itself, that is, refresh itself for new recordings to take place. The inabilty to learn new information is undoubtedly connected with this. Even the ego or persona may occur simply through strong associative interference patterns of past events with present ones.

Finally we come to how the holographic mind "recreates" externalization of experience — the phenomenon of localization mentioned at the beginning of this section. Up to now we have referred to the record made but not to how that record is actually witnessed when "recall" recurs. When a new light beam [NW] passes through the medium, the pattern of waves reaching the eye contains the term

$$[NW] \times [RW]^* \times [IW]$$

If NW happens to be the original RW, this term becomes

$$[RW] \times [RW]^* \times [IW]$$

the [RW]* multiplying [RW] wipes out the reference information in and of itself. This leaves [IW] unadulerated. Thus when the eye sees this term, it is fooled into thinking that it sees [IW] itself, the information wave that was originally coming from the object. Furthermore, it is fooled into thinking that there is a real object "out there" in back of the recorded medium. This is why this image is a virtual image.

If all sensation is recorded in this manner in the brain cortex, then all sensation will be the reconstruction of objects in space and time from apparent or virtual images of those objects recorded in the cortex. All that we sense as "out there" is projected from our "witness" of the recorded virtual images. This would explain Békésy's results and, because our minds are reconstructing images, offer as a scientific hypothesis the basis for the age-old Buddhist-Hindu wisdom that "All is Maya (illusion)."

We don't feel what we feel; we feel what we think we feel. Now repeat this phrase with "smell," "taste," "hear," "see" (and even perhaps "think"). The period of time between the actual event and the movement of the event into the recording cortex is sufficiently small to give us the further temporal illusion that what we sense occurs at the instant the event occurs.

Neurologist Richard Restak argues that we aren't really conscious of events at their occurrence, and consequently so-called free will is an illusion. Our actions in the universe may take place before we even know they have. In a recent article he describes an experiment performed by neurophysiologists Benjamin Libet and Bertram Feinstein at Mount Zion Hospital in San Francisco. He writes:

In one notable experiment, Libet and Feinstein measured the time it took for a touch stimulus on the patient's skin to reach the brain as an electrical signal. The patient was also requested to signal the arrival of the stimulus by pushing a button. Libet and Feinstein found the first detectable electrical signal on the brain's surface occurred in 10 milliseconds (thousandths of a second). The patient's response — the button pushing — took place in one-tenth of a second. But strangely the patient didn't report being consciously aware of the stimulus and response for close to half a second. Remarkably, the data indicated that the patient's conscious actions were somehow referred back in time so that they helped create the comforting delusion that the stimulus preceded action instead of the other way around.

This unconscious action occurring before one is aware of conscious action could be due to two quite different reasons. The first reason is simply that thought awareness (the idea that something has occurred "out there") may only reach our minds after some time has passed — in much the same manner that we move our hands from a hot stove and later feel the burn. In other words, there is a reaction time for conscious awareness of events in the same way that there is for other events. Thought or symbolic word consciousness just takes longer.

But there is another, admittedly far-out, possibility: The signal comes from the future. The conscious awareness occurs as "cause" and sends the message back in time to the response. The patient reconstructed the scenario after the facts in the correct time order — response following stimulus.

Restak points out that a later study discovered that a full 1.5 seconds before a subject decided to lift a finger the brain had already begun to generate waves and signals preparing for the event.

If consciousness takes the form of a qwiff, then to make an event (even as innocuous as lifting a finger) occur, the future thought must "shake hands" across time with the past event. The qwiff from the past moves forward in time to the awareness event and reverberates backward in time as the (complex-conjugate) mirror image of the stimulus event. Together these waves multiply, creating the record of the occurrence — the occurrence and awareness of the occurrence in consciousness. We choose what we have experienced. We create the past, not the other way around.

With the multitude of events simultaneously competing for our "now" attention, and with the understanding that nothing that happens is fundamentally mechanical but that things occur as a result of complementary choices, there must be "free time" for those choices to occur. That time is the future.

Thus the brain hologram acts as a living, consciously choosing medium. It looks backward in time at all of the information waves, [IW]'s, competing for attention, crying out for registration and thereby the desire to become "real." For it is only through registration, or recording, that an event changes from a complex-numbered qwiffian possibility to a self-referenced "reality."

These recorded realities appear now to consciousness as possibilities. For example, we remember the color of our true love's hair. But was it black or perhaps blue-black? Perhaps it had a reddish tinge? Through reinforcement, generated by further recall, we build up a picture of our true love's hair. Black is the color because there are more sequences of events recorded that reconstruct the beautiful black-haired man or woman of our dreams.

Those [IW]'s that become memory records in the future are stored in the holographic cortex as terms in the registration of [IW] + [RW], where the reference qwiff [RW] is generated by the brain. In this way a phase-saving record [RW]* × [IW] is deposited as part of the recording requirement of complex multiplication. The complex-conjugated mirror terms are generated from the future. The past events generate the qwiff itself.

In this manner we reconstruct or recreate the universe in our own images. These images are continually being modified by conscious choices: however, those choices change nothing now. They already have made their changes. They were the creators of our pasts.

There are two other points about the holographic brain that need to be mentioned. One concerns the space in which all this recording goes on and the mechanism of recall. The second concerns the "real" image that is generated by the hologram. What function does it serve? Although we shall return to both of these in the following chapters, I want to point out something about the recall now. The utilization of the real image I will discuss in the final chapter of the book.

How can the brain hold all this information? Some researchers feel that repeated electrical activity generates a change in the number of neurotransmitters present postsynaptically. Experimental work of D. O. Hebb of McGill University, for example, has him surmise that significant synaptic change

164

requires reverberatory circulation of impulses many times around the pattern that is to be "remembered." In this way it is suggested that synaptic efficacy is enhanced.

Karl Lashley of the Yerkes Laboratories of Primate Biology has convincingly argued that the activity of millions of neurons is required in the recall of any memory. Since the capacity of our brains is limited to around ten billion neurons, we find, by dividing ten billion by, say, two million neurons per memory, that we would only be able to store five thousand memories.

Lashley goes on, however, to point out that a single neuron need not necessarily belong to just one engram (a brain memory record). Instead a neuron takes action in many engrams. If so, then this appears to be in agreement with the spreading, or nonlocalized, wave needed for the holographic model. But where is the record, or engram, being made, and how? I suggest that memory is recorded on a molecular or even atomic scale, probably through correlated spin flips of electrons in water molecules. These in turn could influence the synaptic efficacy as reported by researchers. This would set the scale for memories way down to molecular and atomic size and would certainly support the qwiff concept of consciousness I am forwarding since qwiffs determine smaller-mass atomic electronic behavior. (The larger the mass, the smaller is the quantum effect. Consequently if memories are made of macromolecules, cells, or neurons, the mass of these objects is so large that they tend to behave mechanically, following the obvious laws of Newtonian-Maxwellian determinism.)

Granting that memories are atomically recorded, how can anything that small be recalled? The images of the world are macroscopic, not atomic. The answer is again provided by the holographic model. A hologram can also provide a magnified image of its record. If the wavelength of the reference wave used to record the image was ten-billionths of a centimeter, but the wavelength of the reconstructing wave, or new wave, is one centimeter — corresponding to brain waves — the image is magnified by the ratio of reconstructing wavelength to recording wavelength ten billion times. By changing the wavelength of the recalling or reconstructing wave, a zoom-lens effect would appear to be seen by the "observer." Thus an image can be "seen" to be as large as the reconstructing wavelength permits or as big as the brain dome. Further magnification can be made depending on the ratio of the distances of the image plane and the object plane from the hologram.

Thus it appears that the holographic model offers much for the understanding of human memory. To recall a memory one literally ignites a wave pattern in the brain. This in turn creates the experiences of the physical world we live in "out there."

7

New Experience,
Learning, and Intelligence

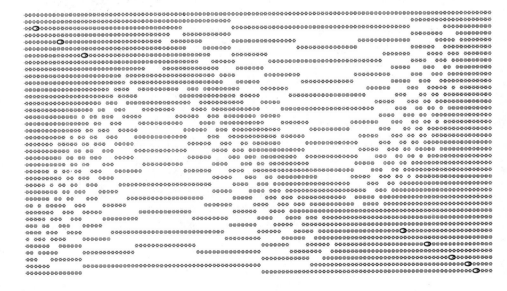

Consequently and paradoxically, in quantum physics one actually gains infor-mation about the future by knowing less about the past.

Quantum Ghosts in the Machine

Through quantum mechanics we possess an altered view of our role in the universe. We have discovered that we each carry with us a theoretical tem-plate, a double-edged ruler that guides our everyday actions. At its most fundamental level this template has two complementary "edges": wave and particle edges we use to measure the momentum or position of a subatomic event. When we use one of the edges, the other becomes inoperative. Through the use of one edge all information that we could have obtained by using the other edge becomes unknown and indeterminate. In this sense a previous order is disrupted by our search for further order. We can compromise, but the compromise never leads to more information, always to less. We must disturb in order to learn.

We learn and order experience through observation. Each act of observa-tion attempts to control and manipulate the universe. We cannot observe without some idea of what it is we try to observe. This thought contains within it the seed of the control and manipulative ability we must have to gain any knowledge at all. Each act depends on how we choose to think about what we think is out there. Our living observations disturb the physical world and thereby increase disorder. We can't help doing it. We can neither avoid nor stop doing it. We cannot become passive observers. We are active even when we think we are passive.

This paradoxical aspect of human learning means that we can never completely understand the universe. The very attempt at understanding cre-ates situations we cannot depend on for certain. Thus any process we believe to be fundamental will turn out to be only one way of seeing the truth, not the truth itself. There is no "truth itself," except the fact that there is no truth itself. While this makes us feel uneasy and even fearful, it also provides the freedom in which new inquiry, learning, and exciting new discoveries can and will be made. It means that what we teach others must always be taught with humility. No one will ever know "the truth, the whole truth, and nothing but the truth."

What is learning? Intelligence and learning make up one of the most perplexing problems facing humankind. Because humans surpass animals in brain volume, it appears reasonable that that is or should be enough to explain human intelligence. However, it can't be the complete answer. Elephants and some other mammals clearly have larger brain volumes than humans. Brain size can't be the determining factor, as important as it is. Maybe brain connec-tions, the number of neural contacts, determine intelligence.

Klopf suggested that human intelligence is based on conditioning brought on by neural pleasure-pain hedonism. Each neuron operates to pursue pleasure and avoid pain.

Kandel and Schwartz found a molecular basis for learning and memory in the simple sea snail *Aplysia*. They measured neural sensitization, showing that it depends on the chemicals serotonin and cyclic adenosine monophosphate (AMP). These two regulate the take-up of phosphorus by nerve proteins and thereby modulate synaptic strength (the ability of the synapse to release transmitter molecules and thus enhance interneuronal firing).

Although both of these approaches are brilliant, each rests on reductionist, neoclassical, Newtonian-mechanical hypotheses. Neither recognizes any role for consciousness as a fundamental "entity" on its own.

In this chapter I would like to offer a fundamental hypothesis concerning the quantum mechanics of learning and memory, based on the creation and destruction of quantum-physical correlations or Einstein-Podolsky-Rosen (EPR) connections. These connections are the occurrences of correlated simultaneous, or spacelike, events in separate regions of the brain.

Briefly, spacelike events are events that take place at different locations but at the same time, so that no signal can travel between them (see Chapter 1 for more discussion). Since these are correlated events, the larger the number of them, the greater the intelligence capacity of the thinking being. Habit is EPR-correlation-creation and intelligence is EPR-correlation-destruction.

These connections can take place only in the human nervous system because of the manner in which symbols, words, and "prethoughts" are correlated there. Furthermore, consciousness is fundamental or axiomatic and not derivable from the materiality in which it occurs. Quantum "ghosts" haunt the machines.

The ability to correlate information in the quantum-physical manner could be a recent quantum jump in evolution that occurred at about the time that humans developed language and the ability to manipulate symbols.

L. Bass, in developing his quantum-mechanical model of the mind-body interaction, suggested that the ability to use conscious direction (thought or "prethought") of muscular activity appeared even earlier than consciousness, yet at a late stage of evolutionary development. His model, a forerunner of mine, requires the evolution of a macroscopic "device" within the neural network capable of correlating data within each neuron in two distinct ways. In the first, or "pure," way data are naively learned. In the second, or "mixed," way data are repeated or already learned. In this distinction Bass postulates an internal "observer" able to tell the difference between the pure and mixed states of its neural correlates. The pure state requires a direct conscious awareness, while the mixed state does not. In the mixed state, the neurons fire automatically, directing the muscles to respond. Thus Bass accounts for autonomic functioning resulting from learning.

In the EPR intelligence model, data are correlated intraneuronally: Firings in one neuron are accompanied by correlated firings in other neurons, even if those latter neurons are on the other side of the brain as much as 10 centimeters away. For this to happen something rather fantastic must arise. This is the origin of how humans come to sense "meaning." I call this "EPR-meaning," the "superconcept." I suggest that meaningful intelligence requires superconception in the mind-brain and that this arises from or is dependent on quantum-physical EPR correlations occurring at the molecular level, probably in water molecules, in our synapses.

Furthermore, a fundamental sense of choice operates at the quantum level of observation. This is the principle of complementarity described above. Herein lies the "force" behind evolution. Consciousness uses complementarity to evolve by choosing how to observe and thereby create the physical universe. Wrong choices result in only one thing — repeated genetic occurrences, i.e., the lizard and the cockroach. Right choices also result in only one thing: All matter becomes consciously aware.

This "matter aware" program of consciousness arises, I believe, through quantum-physical EPR correlations. These correlations enable us to grasp how we experience that "inner sense" of excitement that has made us intelligent. It appears that quantum physics may be giving us a basis for understanding human intelligence and human learning.

The Field of Unlearning

Learning must be exciting, literally. Neurons must fire and will do so only when excited. To learn something, each of us must get "turned on." But suppose we don't. Some people never seem to learn. Do they have so-called learning disabilities? If yes, how were these acquired? Learning did not take place in these individuals with acquired disabilities because it had already taken place.[†] Somewhere along the line of information processing we call learning, the learning-disabled person had already programmed his or her inability to learn. For this to have occurred, the person created a correlation linking the new information being sensed together with the stored thought that the information was of no use.

In other words, only sequences of data that were "useful," or correlated with survival/sexual needs, were retained. Thus "Johnny can't read" because on some level of his mind he doesn't believe it is necessary to read. He can't write because he has the same feeling, and he can't add because "only kooks and weirdos do math." Such disabilities in information processing were acquired through EPR correlations.

[†]Assuming, of course, that the person has the normal ten to fifty billion neurons and no pronounced physical, brain, or nervous system abnormalities.

These correlations, because they are spacelike separated, seem to appear outside of space and time. When they result in repeated behavior, they appear as a physical field.[†] This field generates order "across time" in the sense that a magnetic field generates order across space. This field grows in strength as more and more examples of it are provided. For example, when crystal growing was first invented, it was extremely difficult to create what physicists call single crystals. But through the years, even though the techniques have remained pretty much the same, the success rate has radically improved.

In animal behavior a similar "learning curve" phenomenon takes place. If a number of rats learn a new trick that rats have never performed before, then other rats of the same breed all over the world are able to learn the new trick faster (experiments of William McDougall, Harvard psychologist, 1920; also Crew, Edinburgh University; Agar, Melbourne University — all cited in Sheldrake's book). This occurs even if there is no connection between the rats.

This field is, in other words, a basis for memory that is neomechanical, transcends space-time, and preserves causality. It explains autonomic functioning but does not give a clue to evolution of intelligence. Once an animal acquires the "desired" evolved characteristic, this field explains how others learn it even if those others are vastly separated in space from the originator. It does not explain the origin of creativity or how the new characteristic was created the very first place and time.

A similar field exists between humans. If a certain critical number of people learn to do something well, then it will be far easier for the rest to learn. Similarly, if a number of students connect up the learning process with a negative attitude — "fer-sure zero to the max, I mean a real loser" — then learning becomes more difficult even for those who wish to learn. Tradition is nothing more than the creation of an EPR-correlation field. Religion is also the same. Reactionaryism, particularly a violent one, is the most drastic form of this field. Less extreme are group consciousness, patriotism, and family-hive consciousness.

This process involves a contagious, consciousness-demeaning infection created by the attitudes of those we emulate, such as peers, teachers, presidents, and rock 'n' roll stars. But most likely, in the case of unlearning, it has been created by the minds of the learners linked together into an EPR-correlated field of unlearning.

Nature continually battles with consciousness, saying to it, "Enough already. Stop thinking. It's perfect just as it is. Continue to repeat and all will go well. No need to learn. No need to improve. No need to symbolize and reduce everything to dead and unfeeling equations." And consciousness, lacking materiality, cries out, "You are a fool, Mother Nature. If you could see

[†]In a recent book, biologist Rupert Sheldrake described a similar field, which he called the *morphogenetic field.* I believe that Sheldrake's field is the same as the quantum-physical EPR-correlation field responsible for learning and intelligence.

what I see. If ... only realize yourself beyond the barriers of space-time,
beyond the barriers of time, beyond and beyond. ..." And the pockets of
consciousness locked into stone are long deaf such pleadings. Soon, in eons
and will once again experience
member. Dimly they feel both
up their solidity.

goes on in each atom of us.
ut the quantum principle of
ing the correlation.

om when the teacher is the
g the correlation, shown by
at infectious new "seed" is
the good teacher or guru.

epeatability. This pattern
omplementarity principle
boxes so that evolution
ter conscious.

s of Learning

learning through quan-
anner that no mechani-
puter can duplicate. It
ting entropy-reducing
yway).

s is special in the way
it co... han-light-speed com-
municatio. elation is inherent in
the single-part...

*The arguments of EPR ... ell to certain proper-
ties of a single particle or sys... ..., in analogy with Bell's Theorem, the
predictions of the quantum theory ... are in disagreement with ... any local
theory.*

By "local theory" Franson means any theory that assumes a mechanical,
causal order as the only means for propagating information from one point to
another in space-time. He notes that the ability to form correlations beyond
causal space-time is inherent in the nature of the quantum wave function itself.
It arises as the indeterminate complementarity of every single particle/system
of matter-mind. This EPR correlation, or quantum inheritance, I called in my
earlier book the "Einstein Connection."

The origin of the human being's sense of person could well be acquired or learned as a superconception through these EPR correlations. Such correlations in our quantum brains are so commonplace as to be totally ignored. Further, these correlations form a net of "remembered" electrons — electrons whose correlations link together in a kind of moiré pattern. This pattern gives rise to the holographic medium needed for intelligent memory. This type of memory is not necessary for survival, as for example in the conditioned reflex and classical conditioning achieved through modulation of transmitter release in synapses. Instead it is a memory involving symbol making and language.

Romantic memories are made of EPR correlations between electrons. These memories are "screened" or witnessed as a self-referential process involving photonic qwiffian reconstruction waves (see Chapter 6). When we see ourselves, electrons look at electrons. Through EPR correlation these memories are created with photons acting as the reference beam, outside electronic sensing and other stimuli acting as the information wave, and EPR-correlated electrons acting as the recording medium. In this way two electrons in separated atoms, while maintaining their undefined spin states within those atoms, would nevertheless "know" of each other's presence as an oppositely directed spin. Thus correlation would occur as the record created by the interaction of both the reference beam and the incoming information wave with both electrons simultaneously. By including more electrons in the correlation the memory created would be enhanced.

For example, in a single water molecule the two hydrogen atoms each possess a single electron. However, the spin of that electron can be either up or down with respect to the spin of the other hydrogenous electron.[†] A photon qwiff "creates a record, a miniature bubble" or magnetic domain on the scale of a single H_2O molecule either by correlating a pair or more of electrons or by destroying such a correlation (memory wipeout).

Figure 51(A – C) shows hypothetically how electrons can be used as memory-storage devices. In A we see two molecules spatially separated with randomly oriented spins. There is no correlation between the four electrons. Consequently they have independent qwiffs shown in the figure as ↑1, ↓1, ↑2, or ↓2. Because of the lack of correlation each electron is independent of the others and one can say that each electron "belongs" to its respective hydrogen atom or *hybrid orbital.*

In B the electrons in each molecule have been correlated so that in each molecule the total electron spin is zero but each electron has an indeterminate

---

[†] It is true that the individual "1s" electrons in hydrogen atoms form hydrogen bonds with "2p" electrons in oxygen. These bonds are oriented so that the "2p" electron from oxygen and the "1s" electron from hydrogen have oppositely directed spins. The spins of the individual hydrogen atomic electrons are, however, still undefined. All we know is that each is opposite to a corresponding "2p" oxygen atom electron. There are also four other "hybrid" electrons "loosened up" by the hydrogen-bonding process in water. These are the "2s – 2p" electrons in the second shell of oxygen. The "2p" "hybrid" oxygen electrons are also ideal candidates as memory units and may be more important than the "1s" hydrogen electrons described here.

WATER MOLECULES WITH UNCORRELATED RANDOMLY ORIENTED INDIVIDUAL ELECTRON SPINS

↑ = UP   ↓ = DOWN

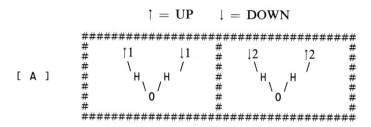

[A]

WATER MOLECULES WITH CORRELATED ELECTRON SPINS ON EACH INDIVIDUAL MOLECULE

[B]

WATER MOLECULES WITH CORRELATED ELECTRON SPINS ON BOTH MOLECULES

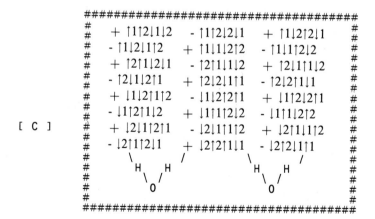

[C]

Figure 51

individual spin. This is shown in each term of the qwiff. For example, the probability that the first electron on molecule 1 has spin up (↑) and the second

173

electron on the same molecule has spin down (↓) is given by the term ↑1↓1. Since there are two such terms, the probability for this is 1/2. Similarly, the probability that the first electron on molecule 2 has spin up (↑) and the second electron on the same molecule has spin down (↓) is given by the term ↑2↓2. Because of the correlation each electron is no longer independent of the other electron on its molecule and one can no longer say that each electron "belongs" to its respective hydrogen atom or hybrid orbital. Instead each electron belongs to its water molecule but is no longer confined to a single atom or orbital. The two electrons on molecule 1 are not correlated with the two electrons on molecule 2. Thus their qwiffs are independent of each other.

In C the electrons in both molecules have been correlated so that in both molecules the total electron spin is zero and each electron has an indeterminate individual spin. For example, the probability that the first electron on molecule 1 has spin up (↑), the second electron on the same molecule has spin down (↓), the first electron on molecule 2 has spin up (↑), and the second electron on the same molecule has spin down (↓) is given by the term ↑1↓1↑2↓2. Since there are twenty-four such terms, the probability for this particular arrangement is 1/24. The two electrons on molecule 1 are correlated with the two electrons on molecule 2. Because of the correlation each electron is no longer independent of the other electrons on either molecule and one can no longer say that each electron "belongs" to its respective water molecule. Instead each electron belongs to the double water molecule complex but is no longer confined to a single hydrogen atom, orbital, or single water molecule. Thus their molecular qwiffs are no longer independent of each other. Instead they form a complex qwiff, taking into consideration all the possible arrangements of four electrons with two possible spins each (↑ or ↓) and two possible locations each (1 or 2).

The "movement" from A to B to C is a dance of correlation in which information is being extracted as the correlation increases. The highest state, C, actually has less information and a smaller entropy than B, which in turn has a smaller entropy than A. When an observation occurs, the C state is suddenly reduced to the A state. This sudden increase in the entropy of the molecular spin system results in the production of information. This production is the experience of learning known as a "mind opening" and often marked by the exclamation "aha."

This frees the correlated electrons from their "correlation traps" (not from their molecules; that would take ionization energy) enabling them to recorrelate through preferential interactions caused by new sensory inputs. These in turn result in memories and reduction of entropy. In this way, through correlation "build-up" and release, life through learning goes on. In a certain sense we are what we learn.

174

The Addiction of Consciousness

One of the first researchers to realize that consciousness cannot be localized was Wilder Penfield. After pursuing the idea that memories were stored locally (for example, smell memories go here, taste memories go there, thoughts go over there), he realized that he was up a blind alley. He stated

In 1951 I had proposed that certain parts of the temporal cortex should be called "memory cortex," and suggested that the neuronal record was located there in the cortex near points at which the stimulating electrode may call forth an experiential response. This was a mistake. . . . The record is not in the cortex.

Where is memory stored? The amazing answer is, nowhere and everywhere. Another way to put this is to say that memory is holographic, with the universe playing the part of the complete hologram. Thus there are records but they are not stored in the computer modular format that many are looking for.

Memory is contained as the stable quantum wave patterns throughout the realm of nature. These patterns are self-reflections reinforcing themselves through trapped consciousness by continued observation, using repeated operational procedures or repeated patterns of complementary observables.

I call this type of reinforcement *consciousness addiction to materiality.* This addiction may be the root cause of all addiction — the fundamental need of matter stabilization through consciousness qwiff-trapping into self-referential, miniature egos. Thus addiction arises out of a need for stabilization — the need to form repetitive patterns. Every particle of matter is its own self-referring qwiff. Thus matter plays the game of Narcissus. It tends to see itself in its own image. Each act of "seeing" is a tiny act of consciousness, a qwiff popping that disrupts the phase relationship of the qwiff with itself but does not change the energy level associated with the particle. It acts much like a disturbed and lonely child who continually hits his head against the wall to make sure that "something" is there. Drug addiction follows a similar pattern. The need is to return to a world of pleasant sensations. The problem is that drug use is taken by the body as a clue to stop manufacturing the natural amount of drug already within the brain's capability to produce.

In this manner the qwiff patterns are not disrupted even though the objects searched for are continually disturbed, thus causing an artificial sense of excitement or pleasure stimulation.

In crystals the stable patterns are not called memory. They are just crystalline structures preserved in energy conservation.

In gases the memories are quickly transformed and frozen as energy-level structures in individual molecules.

In liquids memories can be maintained through long-range order built up by electron correlations on neighboring molecules.

All this occurs in the nervous system at the atomic and molecular level in the water contained within the synapses. I also believe that water acts this way through electron correlations.

These subatomic events occur all of the time at rapid rates. These rates could be dependent on several factors, such as the body temperature. Too little body heat or too much could affect the ability to remain conscious and to learn.

The neural pathways built up and organized into columns in the neocortex appear to be dependent on the release of neurotransmitters in the 300-angstrom-wide synapses (one angstrom is one ten-billionth of a meter). These synapses contain at least 1.2 billion cubic angstroms of volume and, conservatively estimating, about 100 million water molecules. Thus it may be that the columnar "crystals" of the neocortex result from "conscious" water in interaction with sensitive chemical processes such as neurotransmitters release within synapses.

The processes of quantum correlations from different synapses, even widely separated ones, give rise to the personal consciousness. But who or what is this person? Who is the observer of all these events? Is the person to be found in just the causal temporal sequence of such neural events?

Another noted brain researcher, Dr. E. Ramon-Moliner, a physician and brain anatomist with the Department of Anatomy and Cell Biology, Université de Sherbrooke, Canada, believes that consciousness cannot occur as a linear temporal sequence of events. He wrote, in a recent letter to me (based on his earlier article), "[It] was a physical impossibility to write the history of certain brain events in the form of a real time sequence."

By computing the number of neuronal depolarizations taking place over a volume of about 1000 cubic centimeters (the neocortex) to be of the order 10^{12} per second (a trillion times per second), one sees that in the time period of 6×10^{-12} seconds (six-trillionths of a second), which is the time it takes light to travel from the occipital to the frontal pole of the brain, at least three hundred forebrain neurons have undergone depolarization. There simply is no way in which those depolarization events can be causally related. If there is meaning ascribable to these events, how does this meaning take form?

Ramon-Moliner then described the following amusing situation:

Let us assume that a human being was told: "We are going to cut you in two pieces; one of them will be destroyed; the other will be kept functioning as before by artificial and reliable methods; you must tell us which one must be preserved so that your mental (cognitive) functions will continue."

When confronted with such a choice, any human being will give up his legs, his arms, and even most of his body. . . . until reaching his brain. This means that, somehow, we localize ourselves in our brains. But suppose that we continue the experiment. What if my brain is cut in two fragments? In which fragment do I "locate myself"? Having spent a good deal of my life reading about the brain I may give up my cerebellum, a portion of my brain which, supposedly, can be removed without interfering with my thought processes. But what next?

The point I want to make is that, as an entity capable of cognition, I can locate myself in the brain but I cannot go further. I cannot locate myself in any particular nerve cell, let alone a neural transmitter molecule, or an atom within that molecule. It is a sort of "neural uncertainty principle."

The Brain Hologram Revisited

Further on in the letter Ramon-Moliner writes,

I am more inclined to believe in the "holograph" analogy. . . . As you cut an holograph in fragments, each one can be used to reconstitute the totality of the picture, *although with a loss of resolution when the fragments become small. Likewise, I believe that the mental separation between objects of cognition is not linked to spatial separation in the brain.* . . . *the same cognitive act is performed by various points of the brain "at the same time."*

Ramon-Moliner brings us face to face with the "I." He points out that the "I" cannot be localized in any space smaller than the 1000 cubic centimeters of our brain cavity; that to do so leads to logical paradoxes. By bringing out the hologram concept he brings to light that there is a little bit of "I-ness" in every drop of "brain-ness." Yet he balks at the idea that there is any "I-ness" in a neuron.

The "I" or ego concept could arise through EPR-like connections between electrons in the brain, which when correlated produce a surface, or superficial, ego state, as I described in Chapter 6. The more correlates, the stronger the ego. A person with a complete lack of ego would appear as an idiot or possibly an extremely wise person with great mystical/spiritual insight. To see how this occurs, consider how two particles in an EPR-correlating experience acquire identical properties on observation. When two particles fly away from each other after interaction, their properties are potentially acquired at the time of their separation. But the key point here is what those properties are and how and to what extent we can say that they are "objective" properties.

In Chapter 1 I looked at how two particles together compose a state. This state is a maximal description of a physical property that can only be realized by a measurement on both particles at the same time — a measurement that does not test their separateness but a physical property that depends on their togetherness. This is shown in their similar wave-function appearance. This wave function incorporates the uncertainty principle, which does not allow any correlated physical objects to have both pairs of complementary attributes at the same time. Thus each particle has a potential position and potential energy, but each particle does not have an actual position and an actual energy.

But now let the particles separate *and* maintain this state. As long as no one measures anything about each particle, this state is preserved. Now suppose a person measures some property of particle one, say, the energy. Then

177

the state of *both* particles is instantly changed because of their past acquisition. Thus it is also true that each particle acquires its properties at the time of measurement of one of them and they each had these potential properties at the time of their separation (initial interaction).

This acquisition cannot be explained by a mechanical connection. The act of measurement performed on particle one "instantly" reaches particle two. No force field can travel that fast. It is just there.

In our brains this thereness is felt as "I-ness."

Ramon-Moliner's work, if I read it correctly, confirms that there must be EPR correlations occurring in the human brain. Because these correlations do not take place in any unique temporal order, a sensation occurring in one neural pathway would impose some similar "sense" to a neural sensation in another pathway *in a subjectively determined manner.* The brain, acting as an observer, would experience the events as connected and would even attempt to sense a temporal causal order to the events. This would result in something like the famous Pavlovian dog response of salivating to the ringing bell. It is necessary that the correlation be between two or more sensations that could not be cause-effect related. (We call two such events spacelike separated.)

In Figure 52 we see two water molecules separated by several intervening neurons. Each molecule, however, contains a correlated memory created by an earlier interaction possibly caused by a previous molecular collision or interaction with ions involved in the transmission of nerve impulses such as the calcium $2+$ mechanism discussed by Kandel and Schwartz. Even though the molecules are now spacelike separated,[†] an event occurring in neuron 54 will instantly be felt in neuron 1. This event, which I call a qwiff "pop," frees the two water molecules, creates entropy, and is "felt" or identified as a meaningful synchronistic occurrence. This "feeling" is the arising of a personal "I-ness."

If neural events happening in spacelike separated regions of the brain in any sense constitute "knowing," then this knowing cannot be localized; or can it be that any one event is causal with respect to the other? Quantum mechanics through the EPR correlation would indicate that a cognitive act occurring at one place in the brain would be accompanied by the same cognitive act at an entirely different spot — spacelike separated.

These two events (or more, depending on the number involved) happening simultaneously in the brain from one observer's point of view would be interpreted one way. From another observer's viewpoint the same two events would have a different meaning. And from a third, still another meaning. These interpretations thus would constitute a set of roving observers in the brain — possibly ego states or other "I" personalities — each possessing a different perspective. These perspectives may be what we call the person.

[†]Remember that spacelike separated means separated in space but not in time. It is the marking of events simultaneously occurring in different locations.

THE EXPERIENCE OF LEARNING: A TWO-STEP DANCE OF ELECTRONS IN MOLECULES

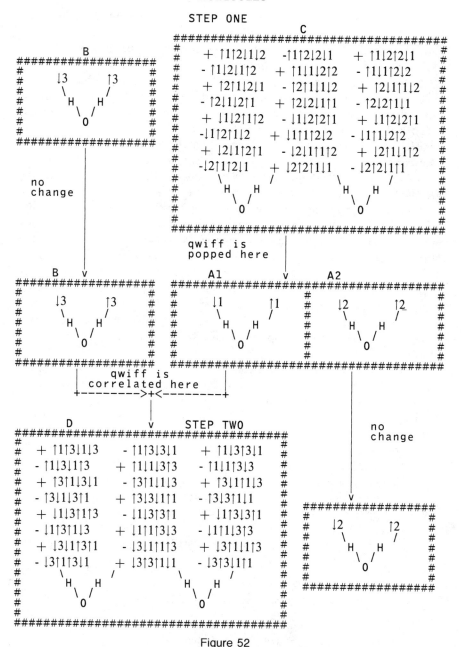

Figure 52

Step one: C → A while B remains the same, causing the release of past information through correlation breaking. Here consciousness enters. Step two: A1+B→D, the storage of new information through the creation of correlation. Here no consciousness is necessary.

One further point about the roving observer in the brain. Since he or she is looking at spacelike separated events, his or her appraisal of their time order will be dependent on the "speed" he or she passes by them. These events could take on different meanings accordingly. Thus a smell could be taken as the cause of an erotic sensation from one point of view while to another the erotic sensation was the cause of the smell. Meaning would arise again when for the first observer (smell $\rightarrow$ eroticism) he or she again felt an erotic feeling and then had the déjà vu of the smell of roses (the time-reversed eroticism $\rightarrow$ smell). Or take the statement "Smoking may be the cause of lung cancer." Reversed, we have the possibility "Lung cancer may be the cause of smoking."

This speed-dependent observer is a scanner of the *brain-states.* Training to think in the usual Western left-brain orientation may simply be one-way scanning. Although highly useful for learning to think logically and verbally, this may be the cause of the idiot savant or fear/ego domination of consciousness. By "shining" the reference holographic wave from the right side of the brain to the left, instead of from left to right as in the manner of Western logical thinking, humans may learn to experience the complementary alogical artistic/mystical vision. The events in our brains may still take place in the same ways that they do, but another observer scanning those events will see them with entirely different meaning. This may be the only difference between a God-realized or cosmically conscious individual and the proverbial "slob in the street."

In this manner we have a mechanism based upon consciousness — qwiff popping in different parts of the brain — which acts as the holographic decoding of stored information. The more correlated pops that occur within a given spacelike separated interval, the more intense is the recall experience. Furthermore, we also have the inverse mechanism for encoding holographically — the unconscious qwiff correlating through interactions governed by the laws of energy and momentum conservation.

In Figure 53 we see the processes of local decoding and encoding, or qwiff popping and qwiff correlating. The latter process is what the Buddhists call *karma.* Karma is energy-conservation memory built up as pockets of trapped consciousness desperately attempting to free itself from material suppression brought on by its own desire to experience itself as matter. In physics it appears simply as the law of conservation of energy.

Cause and Effect and Memory

Although a fool and his money are soon parted, we cannot say the same about a fool and his memories. A person is his memories and in turn a memory is the source of the conservation of energy.

Karma and Freudian cathexis arise as the natural consequences of the law of trapped consciousness. This law in turn appears as the accepted experiences of causality that physically manifests as a statistical average of billions of

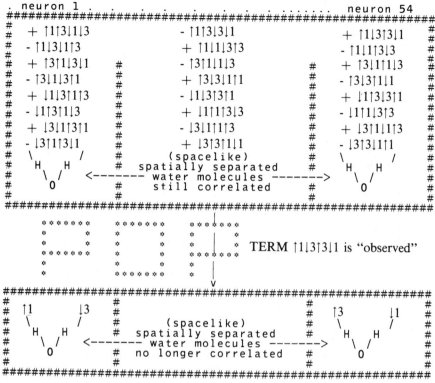

Figure 53

events, known as the law of conservation of energy. All this arises from trapped consciousness or, put another way, is memory.

As we have seen, there is no actual location for memory. Memory is everywhere — in the brain, nervous system, synaptic junctions, muscles, and even in the space around us. No wonder that old houses bring back memories. They are still there!

Memory is holographic. It is inherent in space-time and manifests in the idealistic Newtonian mechanical theory as energy. The ideal ball that rolls down the hill remembers just how far to roll back up the other side. As it rolls it converts perfectly the energy of its gravitational attraction to the entire planet Earth into two forms of kinetic energy: rolling energy and translational energy. Through its distribution of inertia or matter, called and calculated to be its *moment of inertia*, it "knows" just how much energy to turn into rolling and just how much to turn into translating down the hill and up the other side. Its mass distribution is its karma. Equally sized lead balls and aluminum balls, although of quite different weights, all reach the bottom at the same time when rolled down the hill; but an aluminum sphere will beat a lead or aluminum cylinder of equal radius any day.

Through its interaction with the field of gravity the ball's motion and position on the hill is correlated perfectly. There is no entropy, no gain or loss of information in the rolling process, just a transformation from one form of energy to another and back again.

Once there is an established repeatability, the notion of cause and effect arises naturally. We ascribe as cause the early or initial conditions of the ball at the hilltop. We take as the final condition, or the effect, the position and velocity of the ball at hillside or at the bottom. A pattern of repeatability produces a sense of reason. Determinism is nothing more than the connection between a cause and an effect.

This perfected picture of matter in motion does not involve the observer of the perfection. There is nothing to be learned or gained. It is a mindless enterprise.

When looked at quite carefully, however, we see that the ball itself is made up of billions upon billions of space-time events — qwiff poppings of consciousness. Each pop is nearly exactly the same as before. On the time scale of atomic events the pattern of bubblelike pops of consciousness occurring within the ball is hardly changed by the ever-so-weak gravitational field of the earth's hill. And yet these atomic pops are changed. Each atom "feels" the field of the earth. Through this ever-present, ever-changing gravity field each atom is slightly affected. This interaction correlates the earth to the ball's atoms. The probability flow associated with the qwiff *is* the ball. The ball's consciousness is trapped into a simple, classical, inertial response to the macroscopic earth. The ball's inertial mass and its repeatable response is its karma.

Even our molecular genetic code has a karma. Chain molecules of DNA follow a nearly mechanical picture, each behaving as a tiny cause-and-effect machine, programmed to repeat certain cycles endlessly.

But the machine of the ball is not perfect. Even if all frictional effects acting on the ball could be removed, even if the macroscopic ball were aimed and tracked perfectly, it would falter in its motion after a while. I remember how shocked and surprised I was to calculate that the quantum principle of indeterminism, although tiny in its effect on a macroscopic object, multiplies its uncertainty and finally throws the object off track. A Ping-Pong ball dropped from a height of about ten times its diameter onto another Ping-Pong ball held rigid to a table will only bounce on the fixed ball about ten times even if it is aimed as perfectly as possible. The reason is that even a perfect drop contains an uncertainty in the coordinates of the drop point to spoil the bounce.

This indeterminism is the freeing of the qwiff from its trapped condition. No matter how large the object, it eventually will decay. The molecules of DNA are no exception. Eventually the qwiffian spirit that inhabits and is trapped by our physical body form is freed. The DNA molecules eventually make an inevitable error. Thus karma is a temporary condition, a repeatability arising from the conditions necessary to produce memory — the conditions of correlations built up through the interaction of fields.

Yet without this error in memory, without this imperfection and demand that consciousness play the roll of qwiff popping, nothing new could be learned by the DNA molecule. It could not alter its performance and make a foot instead of a hand. To learn is to alter radically its own self-serving ego behavior. To form correlations of certain neural pathways in the brain with early conditions of childhood is to create cathexis. The water molecules program the DNA through selected neural synapses. The DNA, already programmed by the primal sea that gave birth to life in that violent time we call the Archaeozoic Age (about three billion years ago), when the trapped consciousness in that sea cried for something to interact with, is once again shocked by the electrical charge it receives when a neuron fires. It resists, holding on to its pattern of repeatable, cyclical, wavelike consciousness, its quantum wave function.

But the electrical charge of a neuron firing is felt within a synapse. Like the lightning bolt striking an Archaeozoic sea, the electrical fire disturbs the DNA, setting up the possibility to correlate its dance with it. To correlate its molecular arrangement with a fiery dance of light, which each atom and each electron dimly remembers, is to face fear and loss of molecular ego. It is a dance of death.

Probabilities are altered in the process. The presence of more electrical fires in the synapse alters the pattern slightly, like the inert ball rolling down the hill, and the pattern radically changes. The next time the neuron fires this way, the dance is autonomically correlated. The DNA dances with light and says to itself, "This is normal."[†] The DNA has forgotten its earlier "self" and the death of that self.

And a thought is made matter. A personality trait appears that is not the simple endless repeat of that of the parents or the parents' parents, or the parents' parents' parents. Evolution, for better or for worse, has occurred. Consciousness has eked out a rung on the ladder to the stars from which it came and to which it ultimately will return.

[†]Some readers may be bothered by my extensive use of anthropomorphisms. How can an electron "remember"? How can DNA "dance" with light? I believe that the words we use, "remember" and "dance," arise from these fundamental interactions at the levels of electrons and molecules. Thus to "dance" with light is the continual process of rhythmic interaction of electrons with photons. An electron's memory is contained in its $E = Mc^2$ equation — i.e., in its inertia.

8

New Laws of Psychology

When foot feels floor, foot feels foot.

The Buddha.

The Connection of Mind and Matter

The most fundamental tenet of physics is that a real physical world exists. This physical world exists, has existed, and will continue to exist irrespective of my own existence as a being. It existed 100 years ago, 1,000 years ago, and probably 300 billion years ago. It existed at the time of the big bang and will exist at the time of the "big crunch," when all matter will gravitationally collapse, one planet onto another and then onto the sun and all suns onto each other, in a gigantic, time-stopping, imploding black hole.

This tenet of objective physicality is held as a sacred axiom by nearly all scientists. They believe that science can be based only on such a tenet and that any statement to the contrary would be a sacrilege to science and thereby rejected as mere mysticism.

Even psychologists and psychiatrists believe that there is a fundamental material basis for mental phenomena — that mind is a catchall concept that includes physical processes too complex to be understood as mechanics.

Faced with this fundamental prejudice, which in spite of its non-provable-ness has served our technosociety well, psychology has been backed into a corner from which it can never return. The Newtonian paint on the floor will never dry. Our Western attitude is just as much mumbo jumbo, insofar as we never question this tenet, as any Hollywood witch doctor's chant.

There simply is no hope for such psychology today. Why? Because this sacred tenet of science is plain nonsense. There is not the slightest shred of evidence that proves the existence of a physical world acting independently of human thought. As ridiculous as this may at first appear, consider that any "thing" you know to be a "thing" comes through human knowledge, human sensation, perception, and conception. That comfortable chair that "sits there" waiting for your return, welcoming you with its feeling of warmth and cozi-ness, is your memory of that chair returning to you. Your sensations of "chairness" are your own. You feel yourself participating in your experience of sitting in "chairness." You don't feel the chair; you feel your rear end. The chair is just an extension of your butt.

How can this be in the light of everything we know today, particularly in the world of the "new physics"? Why, it doesn't even make sense. The chair I am sitting in must surely be there beneath my posterior. I know it is because I can feel "it." I can see "it." I can even smell or taste "it."

As sensible as this may seem to us, the chair is not really there in any sense as a physical object independent of our sensing/sitting on it. If our thoughts about the chair suggest that it exists independent of us sitters, we are in for a shock. The chair has no existence independent of us. Sitters and chairs go

185

together like the proverbial horse and carriage: Each really defines the other.

The reason for this is quantum physical — or, if you like, mind and matter go together like horse and carriage; one defines the other. In this chapter the new psychological basis for human experience will be spelled out. In the "new physics" we have learned that acts of observation create the sensations we call *experiences of the real world* — sensations of that imaginary solid-looking world we take for granted or, as a young child told me, "take for granite," as being "out there." A nearly countless but nevertheless finite number of acts of sensation "make up" the linear sequences we unthinkingly take as our temporal conceptions of the physical world "out there."

Without these minuscule but finite acts of observation we would have no knowledge of the physicality of the world. And, even more, that physicality would not be what we perceive it to be. It depends on us for its existence and we in turn depend on it. The chair depends on me and I depend on it. The house depends on me and I depend on it. The arm in my shoulder depends on me and I depend on it. Each and every object in the universe of my experience depends on me and I depend on it. How?

This interdependence defines the separation called "I/IT." Our knowing that it exists is directly linked through consciousness to our senses that it is there.

For example, a quasar (Q) many thousands of light-years away sends its light to a receptor on earth. That receptor consists of a pair of light sensors. Each sensor is directionally sensitive. The light comes to the receptor via two distinct paths. These paths themselves are over 50,000 light-years apart. The first path is more or less direct. The second path is bent by the gravitational-lens effect of an intervening lensing galaxy, called G-1 by astronomers.

According to the quantum theory of light, the quasar emits photons. Each photon, although potentially only a single event in the recording receptor of the observer, either spreads out following both paths — if the observer does not attempt to record by which path the photon traveled — or the photon does not spread out and instead follows one path or the other — if the observer attempts to record by which path the photon traveled. In the first case the two paths are said to interfere with each other, the interference pattern produced by the setup of the sensing instrument resulting in one of the two sensors experiencing more events than the other. On the other hand, if the experimenter sets up a complementary experiment designed to check which path the photon followed, the photon will be "observed" to have followed one path or the other, giving to each sensor a 50 percent detection probability.

It is the setup of the observer that decides either by how much phase difference there shall be between the two paths (i.e., the photon came by two paths) or by which path the photon came. In the first case one sensor receives perhaps 100 percent of the photons while the second sensor receives nothing. The phase interference of the two light paths cancels out the flow of each path into the second sensor. The interference proved that the 50,000-light-year

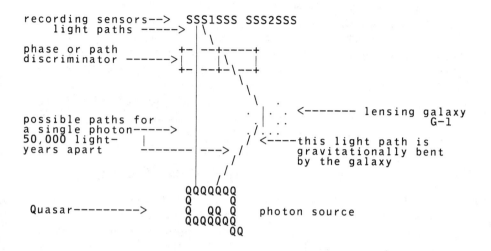

Figure 54

photon qwiff was spread over the two paths at the instant the observer chose to determine the phase rather than the path. In the second case both sensors receive photons with equal probabilities. The path by which the photon came many thousands of years ago was also instantly decided by the observer's choice to measure path.

Such a "delayed choice" experiment was first thought of by physicist John A. Wheeler, who perhaps more than any other physicist has pointed out to us that the world we sense does not exist without our sensing of it. Our sensing depends on our thoughts. As I pointed out in the previous chapter, these thoughts *now* influence the past. That past is what we say is logically recorded as the past according to a sense making causality the preserving view of the world. The history of a single photon, including its arduous trek around a path separation 50,000 light-years wide, was not written 50,000 years ago. It wasn't even written one year ago. It was written, recorded as truth, now, at the instant that a human mind realized it after the performance of an experiment. Our thoughts, although coming much, much later, are not after the fact at all. They are the creators of the fact. We create history now.

From all these considerations many physicists like myself are drawn to make the rash observation that mind cannot be separated from matter. Mind is able, like any good time machine, to reach back into time. But mind does not observe the past; it creates it. The past is only as real as we make it. If this is so for the past, what can we say about the present? The present depends on

187

which choices were made (rather, will be made) in the future. I am as I am because I will be as I will be. I was as I was (in the past) because I am as I am (in the present), not the other way around.

Looking at experience this way means that we can no longer accept our plight as entirely due to our poor parents' genes. (In the classical cause-effect view we can only see cause in the past, arising from the present point of view of effect. We attribute to the past, and to the past only, the domain of cause. We decide what the causes were in order to make sense of our being at effect now. As Werner Erhard, the creator of the est training put it, "We steer our cars through life by turning the rearview mirror."

The Old Laws of Psychology and the "Old Physics"

Comforting though it may be, it is a myth that the past decides the present. All classical physical laws are nothing more than plausibility arguments of enormous persuasion. We use them to convince ourselves resoundingly that it is safe to fly, drive, go to the moon, or to the bathroom. We attempt to use them as a basis for dieting, bridge building, rocketship landing, missile trajectories, and nearly everything we observe, living or dead. Events that in and of themselves could not occur, no matter how likely their occurrence, do occur because they are sensed to occur. For example, imagine that photon emitted by a quasar thousands of years ago. It reaches our eye by one of two possible means: It either travels along two paths simultaneously or it travels along either of the two paths. The single act of sensing, together with the setup designated by the experimenter, decides in the present what sequences of events the photon had to have followed to reach the eye today. This sensing is *the* reason and purpose of consciousness. Without this sensing or qwiff popping no event would ever make it as an event. It would remain invisible qwiffness, only a pattern of probability in space. But when observation, the role of consciousness, enters, it enters with great power in humans, with less power in animals, with even less power in plants, and — going down the line — with minuscule, nearly unconscious, power in atoms, electrons, protons, quarks, gluons. Everything is created by an act of consciousness.[†]

Before we look at the laws of the new psychology, which are based on quantum physics and acts of consciousness, it is useful to review the old psychology's basis in classical physics.

If it is true that the past cannot be the causal basis for the present, then what determines the pattern of events that we each experience as the present? Furthermore, how does the illusion of causality arise? After all, it is this illusion that gives us the reassuring sense of identity. It provides the conceptual framework on which human behavior is based. The old psychology is based

---

[†]Each consciousness act is accompanied by an interaction, but each interaction is not necessarily a consciousness act (see Chapter 5).

on this illusory past that supposedly causes the present.

This assurance arises from classical mechanics, which in turn is based on the assumption that events occur without our consciousness. These events are natural sequences following a temporal order. We see this order when we look at such sequences and observe certain symmetries and similarities. The ball thrown in the air and then caught is matched to the orbiting satellite. These sequential patterns appear to follow a law — the law of prediction. Without prediction all science would come to a standstill.

Predictions come true in classical mechanics. There are no errors. From an initial condition in which the positions and momenta of all objects under inspection are specified, a final condition of the positions and momenta of these objects is predicted. This final condition is predicted to occur with 100 percent certainty. Any observed deviation is considered to result from an error, either in the specification of the initial condition or the influences or forces operating during the trajectories of the objects from their initial condition to their final condition.

Behaviorism, operant conditioning, and reductionism go hand in hand with this classical predictive assumption. Memories act as initial conditions stored within the nervous system. When a novel situation is confronted, forces arise within the nervous system that, when based upon the similarity of the novel event to the stored initial conditions, determine the behavior of the individual. The individual simply repeats the pattern of behavior, following the influences or forces in existence. Thus, according to behaviorism, the outcome is determined because the stored initial conditions must lead to that outcome as the final condition.

Abnormal behavior or inappropriate behavior is a considered result of error, often attributed to an *unconscious* set of initial conditions. Since these hidden initial conditions are thought to be causatively operating, they will produce a predictable outcome. Psychologists fail to make those predictions when they cannot see all the initial conditions. Once the relationship with the patient is established so that these hidden initial conditions, such as a cold father or a demeaning mother, are unmasked, the behavior of the individual becomes "understandable."†

Behavior modification results from this when these unmasked earlier conditions are recognized. Since the person is now *conscious* of them (meaning that a reinforcement conditioning has been applied in which new forces, such as reward and punishment, are brought to bear), the person cannot behave in the same manner any longer even if he wishes to.

The person behaves differently because new forces have been introduced to compensate for the unmasked initial conditions, which, because they are

---

†An important distinction is made in the philosophy of science between explainable versus predictable. Human behavior may be explainable in terms of initial conditions, yet remain unpredictable. Thus we often settle for the *a posteriori* explainable.

"facts," can never be erased or changed. If further abnormalities are encountered or the person reverts to the previous abnormality, more forces are brought in to alter the behavior. These forces include drug therapies and other body-invading devices. In the old psychology, the person is always a victim to his initial conditions and new forces introduced into his system.

Freudian psychology, briefly, attempts to unmask the unconscious as much as is necessary to create appropriate behavior. Freudian forces, like Newtonian gravity and Maxwellian electromagnetism, must be known. Then the behavior is predictable. Although I have not attempted to describe all of the profound insights discovered by the great Western "fathers" of modern psychology, I feel confident that, once recognized, psychologists, analysts, and counselors will see that modern psychology today is based on guidelines provided by fundamental physical science that still follows the Newtonian framework for all existence.

Extending the "New Physics" into Psychology

Here we shall see how the mind is connected to matter, how the "I" arises as a quantum-mechanical wave flowing toward the future and backward in time to the present. Somehow a random element is introduced into the pattern. The mind projects experience and experience projects the mind. Without this double-flowed projection there is no experience, that is, there is no perceived experience.

We could call this the physics of consciousness — how we create the "out there" experience from our "in here" experience. There are laws to this game of physics and consciousness. The fundamental axiom of my research is that the scientific basis for all consciousness is contained in the three laws of quantum mechanics.

But before we look at them we will need to examine certain concepts and their symbolic notation. These concepts are used to describe the quantum physics of atomic particles. The notation for them was first written by Paul Dirac and is found in his book *Quantum Mechanics*. With this notation we shall use the laws of quantum mechanics as the basis for consciousness. The concepts we wish to understand are

- the concept of "state" and
- the idea of the quantum-mechanical probability amplitude

The Concept of State

A state in quantum mechanics is an abstract concept. It represents a condition, as in a state of existence. For example, a physical object can be imagined to be moving like a car along a road from one state to another. This state could be a physical location, such as the state of Kansas. Or it could be a condition of motion, such as 60 miles per hour north on Highway 5. Thus

190

a state depends on some reference base, such as location in space or occurrence in time or both.

To denote a state we write it as "$|A>$." In quantum physics this is called a *ket* and is read ket-*A*. It stands for the idea of a physical state, one that we say could be in existence. It also stands for the idea of a mental state, a choice that can be realized. Thus if *A* is a car moving along the highway at 60 miles per hour, ket-*A* is the *state* of a car moving along the highway at 60 miles per hour. Ket-*A* is to *A* as an idea is to a reality regardless of whether that reality is physical or mental. Since the three laws of quantum mechanics depend on arithmetic laws of ket addition and multiplication, we need to look at them next.

Ket Addition

Kets can be added. For example, it is possible to add the ket

$|I$ am moving along the highway north at 60 miles per hour$>$

to the ket

$|I$ am moving along the highway north at 30 miles per hour$>$

By addition we arrive at the ket

$|I$ am moving along the highway north at 90 miles per hour$>$

It is not possible to really *be* moving along the highway at both speeds at the same time. It is possible to think the idea that one is moving along at both speeds (60 mph + 30 mph) and that the result of this is the movement at a single speed (90 mph).

It is even possible to add two kets that are contradictory. For example, add the ket

$|I$ am moving along the highway north at 90 miles per hour$>$

to the ket

$|I$ am moving along the highway south at 90 miles per hour$>$.

The result is I am not moving at all, or in ket language,

$|I$ am moving along the highway at zero miles per hour$>$.

Again, it is easy to think contradictory thoughts or have contradictory ideas. The result of those ideas could be *real* even though the basis for arriving at that reality was nonsense. Through the addition of kets new things are created. Kets form the mathematical basis for the mind creating.

Ket Multiplication

Kets can also be multiplied, but not in the usual manner. What would it mean to multiply the ket

|*I* am moving along the highway south at 90 miles per hour>

and the ket

|*I* am moving along the highway north at 60 miles per hour>?

It does have a meaning, but to grasp it we need to look at another notation for a state.

It is written as "<*A*|" and is called a *bra*. It is read as *A*-bra. It also stands for the idea of a physical or mental state, one that we say could be in existence. Also, *A*-bra is to *A* as an idea is to a reality. But there is a subtle difference between a bra and a ket.

This difference lies in how we view cause and effect in quantum mechanics. Normally, as in classical, or Newtonian, mechanics, we think of cause as an initial event — something that happens first. We then think of effect as the result of cause — something that occurs later — a final event. In quantum mechanics we think of a ket as an initial state, a probable cause, and a bra as a final state, a probable effect. In the "new psychology" we think of a ket as an initial idea and a bra as a final idea. There is a sense of "flow," or probable causality, between a bra and a ket. Something flows *from* a ket to a bra. A bra can be thought of as a state that could happen after an intial ket occurs. Symbolically we represent their relationship as a product of a bra and a ket. (Think of the word *bracket* as divided into two parts corresponding to the bra and the ket.)

Their product is written <*B*|*A*> and is read "*B*-bracket-*A*." The meaning of *B*-bracket-*A* involves a question: If *A* is true, how likely is *B* true? Or, put another way: If *A*, then *B*? The question mark is important. It implies doubt or *fundamental uncertainty*. It arises from the principle of uncertainty within the quantum nature of the universe itself. If the question mark were not present, the statement would be: If *A*, then *B*. This would imply certainty about the *B* state *if* we were certain about the *A* state. The bracket <*B*|*A*> is not meant to imply such certainty. If such certainty were present, then we would have a causal relationship between *A* and *B*. In quantum mechanics we know that no such causality is possible with absolute certainty. Thus <*B*|*A*> is a question, not an actuality; it is an idea, not a reality. It isn't even a probability yet. It becomes a probability after it is multiplied by its complex conjugate (see Chapter 6). In a sense it is a structure within the unconscious — a question or thought that is yet to occur, to pop into our awareness. It resembles a Jungian archetype and it is called the quantum-mechanical probability amplitude, or *quamp*.

The Quantum-Mechanical Probability Amplitude

In Chapter 1 I wrote about the qwiff, which stands for quantum wave function. It represents a wave form in space and time and is pictured as a wave created by a stone dropping in a still pond and spreading out over the water.

A qwiff is a form of *B*-bracket-*A*, $<B|A>$, the quantum-mechanical probability amplitude. In a qwiff, *A* is the initial event — the stone hitting the surface of the still water — and *B* is the final event — the circular wave spread over the pond. Since *B* represents the physical picture of a circular wave, it utilizes the external frame of reference of space and time within which the wave is thought to exist.

However, $<B|A>$ need not be an idea that necessarily exists in space and time. It could be a product of states of consciousness. *B* and *A* could be abstract ideas of mental states such as feelings or emotions. A qwiff will not adequately describe feelings because they don't exist in space and time. To describe both mental and physical states we need quantum-mechanical probability amplitudes, or *quamps*.

A quamp doesn't necessarily represent physical events in space and time. It can be thought of as an "underground" flowing stream that starts at *A* and ends at *B* within the unconscious. It is the "prethought" form reaching out for materialization in the brain.

But how does something enter our consciousness? In order to understand this we will look at an example in the physical world, for the way that the unconscious $<B|A>$ becomes conscious is the same way that a physical event is observed to occur.

Consider two events, two happenings. They could be the ringing of a telephone and the perception of that ringing. Call the ringing phone event *A,* and the perception of the ringing phone event *B.* Again we envision this ideally in quantum physics as a quantum wave amplitude and write it as $<B|A>$.

The ket event, *A,* the ringing of the phone, is thought of as the initial event, and the bra event, *B,* the hearing of the ring, as the final event. The bracket $<B|A>$ represents an unconscious "psychic" kind of question: Given that event *A* has occurred, will event *B* occur? That is, will it register on our consciousness as a legitimate event? Event *B* is in question. Remember that in quantum mechanics we are dealing with quantum-mechanical probability amplitudes, possibilities, likelihoods, and not actually occurring certainties.

What is about to happen, however, is the *experience* of such a certainty in the mind of the observer. He will hear the phone ring; he will have the sensation of the living experience of that ring. What I am describing is how that process takes place. What occurs now is something strange and wonderful in our "mind's ear," not the ear that hears the phone and registers the event. For the real event to register in our consciousness, our "mind's ear" projects the hearing of the phone ring backward in time toward the actual ringing phone. This projection is unconscious. It is a quamp composed of a bra and a ket, but with the sequence reversed. It is written "$<A\,|\,B>$."

Event *B*, in our minds, generates its own backward-in-time-flowing quamp: From the initial quamp, $<B\,|\,A>$, the reversed quamp, $<A\,|\,B>$, is generated and flows back in time.[†] The hearing ear and the ringing phone "shake hands." They agree to relate. If there were other listeners in the room at the instant you heard a telephone ring, no other ear would have heard the ring that you heard. (Although this idea may sound fantastic, it is used in modern physics. Physicist John G. Cramer wrote about this "advanced potential" in dealing with electromagnetic waves as a way of resolving the EPR quantum-mechanical paradox discussed in Chapter 1.)

Of course, in the next instant another person would hear the ring; and in the next, another; and so on. There would be so many rings per instant of time that eventually all would recognize the source of the ringing. But each recognition that you have is yours. No one else in the room "reads" that ring at exactly the same time you do.

An external event sends its quamp to our brains and our brains answer by sending back into time our projection of that event. Our minds turn our brains into a time mirror. They project the perception of the event back onto the event itself. The process goes *A*-bracket-*B*-bracket-*A*, $<A\,|\,B><B\,|\,A>$, from *A* to *B* and then from *B* back to *A* again. It is a self-referential process in which the physical event *A* and the mental event *B* connect. That is why we hear the phone at the phone. That is why external events appear external. Our minds are not in our heads, although that is where the time-reversed process begins. Our minds are "out there."

This quantum-mechanical probability amplitude, $<A\,|\,B>$, *is* the given hearing of the phone and the probable ringing of the phone. It is the question: If *B*, then *A?* If I heard the phone, then the phone rang? Then the multiplication of the two "questions" parts produces a probable answer, the relative probability $<A\,|\,B><B\,|\,A>$. This becomes a verification, a "shake hands" across time that the phone did ring, event *A* occurred — and I heard it ring, event *B*. Looking at the formula, $<A\,|\,B><B\,|\,A>$, and starting to read it from right to left, like Hebrew, we have an almost liturgical recitation: The phone has rung. Have I heard it? I have heard it. The phone has rung?

Although two wrongs do not a right make, two questions do a probable reality make. Now the two events are connected logically and seemingly causally. Out of the original question and its time-reversed question a reenhanced answer is heard. The phone has rung and I have heard it. I had to project my "hearing" "out there" to the phone back in time when the phone rang. My mind is a time machine.

[†]The atemporal quality of quantum physics' $<A\,|\,B>$ acting in the mind can be related to Sigmund Freud's concept of the atemporal id.

In our "mind's ear," in our unconscious minds, the ket, or initial state, is the hearing of the phone. The bra, or final state, is the phone ringing. To our unconscious minds there is no logical problem. It is OK that first I hear the phone and next it rings. Our unconscious minds, like the Freudian id, operate beyond space and time. As I pointed out in *Taking the Quantum Leap,* quantum waves are not restricted by causality. Quantum physics does not imply a temporal order. The quantum wave representing the simplest movement of a particle from here to there moves with a phase velocity faster than light. Quamps are also not restricted by causality. They are simply questions. To our unconscious minds time-reversed events are normal; to our "common" senses they are weird and constitute a form of psychic phenomena. Yet according to this quantum-physically based model, there is no normal reality without a psychic reality. To hear the phone ring you have to know ahead of time what it sounds like. Yet you don't really remember it at all. It is not on your mind. It occurs to you when the phone rings.

How does that happen? How do we sort out the data bombarding our senses? That is, how do we make sense of the world we experience? The answer is that we psychically manipulate what we experience. And since this manipulation is quantum mechanical in origin, we cannot be sure that its operations are true. We cannot be sure of our data processing, our interpretations of data. We intrepret what we see "out there"; we operate on the flow creating disharmony. Thus our backward-through-time projection doesn't quite match the flow from the "out there" event. The "in here" flow projected back in time is slightly flawed. When it reaches the "out there" event, the flows interact. That means that their waves multiply and thus produce a dynamic interference that cannot be described as a continuous linear mechanical process. That's why I call it a qwiff pop. This popping type of interference, somewhat like the old radio static we heard with toy "whisker" radios, is the consciousness experience.

It is a paradox that in order to experience, in order to have living experiences, we must be imperfect machines. It is our lack of perfection that makes us "perfect," that is, gives us the living experience we call life.

Life and mind are therefore connected. How each living creature experiences life depends on the creature's mind. But what occurs to my mind? How does it know to create the thought of an occurrence, an experience? Somehow I influence the events of my life. I learn things. I will my body to move, etc. In other words, there are manipulative operations I perform that allow me to know that an occurrence has occurred. To manipulate anything we must interpret the world we live in, including our own bodies. These actions of ours somehow influence or modify the "flow." They restrict or alter the quamp $<B|A>$. Something gets between the bra and the ket. In quantum mechanics this "in-betweenness" is called an *operator*.

Why do we alter the world at all? Think of a ket as an initial idea, a concept that we fail to recognize clearly. Again let us symbolize it by $|A>$. Let us imagine that in our "mind's eye" we see an ideal concept, one that we do recognize clearly. Suppose we symbolize it by the bra $<B|$. How can we create the ideal world B working with the unclear idea A? Let us imagine that there exists a B operator. We symbolize it by writing it as $|B|$ and call it *be-op*. An A operator we would call *a-op*. (Can you dig it, man?) The two vertical lines are important. They tell us that $|B|$ is an operator. Now we apply $|B|$ to $|A>$. In words, we "be-op a ket-A."

We interpret what we see "out there" by operating on the flow, creating a disruption of the ket-A. The ket-A is no longer the same. Thus our backward-through-time projection doesn't quite match the flow from the "out there" event. The "in here" flow is projected back in time slightly flawed. When it reaches the "out there" event, the flows interact. That means that their waves multiply and thus produce popping interference. That interference is what we experience as a new experience.

To grasp the physics of living experience we need next to examine its laws.

The Laws of the Physics of Consciousness

In words, the first law of consciousness comes from the first law of quantum-mechanical probability amplitudes. It is:

LAW I. THE MIND IS ONE.

There is one mind and one mind only. We symbolize this law as:

$$\text{I.} \quad <I|I> = 1$$

This is the "I" experience of the one mind. We read this as I-bra, ket-I. It is a double flow producing itself. Yet nothing is experienced, for the one mind cannot experience itself. No experience is possible without an external object. There must be a separation of "I" from something, the "not-I." Here "I" is not interfered with. It is perfected onto itself. It is one, the big ONE, the only ONE that is.

In order for the one mind to know, it must project itself, and that is the second law. That is the second law.

LAW II. THE MIND IS MANY.

From the one mind all minds and all experiences are projected. This means that within unity there is infinity. We symbolize this law as:

$$<I|I> = \text{SUM (over all } i, \text{ from } i = 1 \text{ to } i = \text{ infinity) of}$$
$$<I|i><i|I>$$

SUM means that we add up terms. If we were to write down these terms and start adding, they would look like the following:

$$<I|1><1|I> \; + \; <I|2><2|I> \; + \; <I|3><3|I> \; + \; <I|4><4|I>$$
$$+ \; <I|5><5|I> \; + \; <I|6><6|I> \; + \; <I|7><7|I> \; \ldots \text{ (and so on)}$$

Each term is read *I*-bracket-*i*-bracket-*I*. For example, $<I|6><6|I>$ is read *I*-bracket-6-bracket-*I*. Each term is part of the whole unity. Since each term is double-flowed, each term can be thought of as a real event.

In quantum mechanics this is called the law of completeness. It says that unity can be broken up into separate parts. These parts, composed of the *i*-bras, $<i|$, and ket-*i*'s, $|i>$, are limited states of conscious experience. They could be energy states in an atom or position states of an object moving from place to place. Law II tells us that we can fully represent the universe only by counting all of the infinite states possible. Since we cannot do this in practice we cannot ever completely know the world in terms of its countless ministates, the ket-*i*'s.

Each term in the sum is a conscious experience, and together they compose ALL that is conscious of that particular "state" experience. But Law II and Law I go hand in hand. Together they tell us that this sum must be unity, the number 1. In other words, all conscious experience added together would end up with no distinctions! The one mind would be itself once again, but without experiencing itself.

Another way to look at Law II is to realize that *I* is all of its projections $|i>$. Law II states that any state *I* can be seen as a sum of other states *i*. Just what *I* is and what the *i*'s are really depends on the kinds of questions you ask of nature. *I* can be the state of the whole universe or the "mind of God." But this, although poetic, leaves little for experimentation. *I* could be a particular "soul" or personality state, with the *i*'s being projections or subpersonalities.

In a marital relationship *I* could be the marriage personality exhibited by the couple, with the *i*'s representing distinctions shown by the individuals in their separate interactions in the world. In this case no single personality is seen. In the proverbial bar scene, the woman being picked up by the man notes, "I thought you were married," after a brief conversation. "How can you tell?" the man asks. "I don't know, I just can," the woman responds.

In a nation or city *I* could be the "soul of the city" or the "heart of the nation." The *i*'s could then be the individuals acting within the framework of that "soul." A typical example is seen in the so-called corporate identity. The individual who possesses it deals with the world as a representative of the corporation. The *I* is the corporate personality while the individual in his Brooks Brothers suit is the *i*. The film *The Man in the Grey Flannel Suit* exhibited this quality. When the individual took off his suit he changed his *I*.

If we think of I as a source of light and the little i-kets as its projections through all of its infinite possibilities, we can write Law II in the form:

LAW IIa. "I" AM ALL MY PROJECTIONS.

$|I> =$ SUM (over all i, from $i = 1$ to $i = $ infinity) of
$$|i><i|I>$$

If we were to write down these terms and start adding, they would look like the following:

$$|I> = |1><1|I> + |2><2|I> + |3><3|I> + |4><4|I>$$
$$+ |5><5|I> + |6><6|I> + |7><7|I> + \ldots \text{(and so on)}$$

Each term is read ket-i-bracket-I. For example, $|6><6|I>$ is read 6-bracket-I. Each term is part of the I.

Another way to write Law II is to get rid of the I altogether and realize that Law Ia states that all projections are unity. All is one. We could write Law II as follows:

LAW IIb. ALL IS ONE.

$1 = $ SUM (over all i, from $i = 1$ to $i = $ infinity) of $|i><i|$

If we were to write down these terms and start adding, they would look like the following:

$$1 = |1><1| + |2><2| + |3><3| + |4><4| + |5><5|$$
$$+ |6><6| + |7><7| + \ldots \text{(and so on)}$$

Each term is read ket-i-bra. For example, $|6><6|$ is read ket-6-bra. Each term is a projection of the whole unity.

In order for the one mind to know itself it cannot be so perfect. It cannot sum up to infinity, not if it wants to know anything at all. And therein lies the so-called illusion of the mystics. To know is to disrupt. To know is to cause confusion. To try for perfection is to create imperfection. Law II leaves open the nature of the i's. The important thing here is that all possible i's are included. Each human being contains a multitude of i persona. Depending on what I the person "wears," his behavior in the world is altered. The same base states, or base personalities, i's appear quite differently when they are bracketed by different i's, as in the example of the individual who becomes a corporate I in the morning and a married I in the evening.

So how does mind do it? It "ops." It uses operators, $|A|$, A-ops, for example, to create from the "big $|I>$ in the sky" the "useful" ket-A states, $|A>$. To do this we need the third law of the physics of consciousness.

LAW III. THE MIND CREATES.

To each and every observable there is a psychic operator. For example, suppose that A is "a car moving at ____ miles per hour." Of course, to describe the moving car more accurately we need to fill in the blank. Since cars can move with many different speeds, we let each different speed be symbolized by the letter i. We then write Ai to mean "a car moving at i miles per hour." And we symbolize this as ket-Ai, $|Ai>$, a particular state of movement of the car. Now we bring in A-op, $|A|$, which is the operator that creates from ket-Ai, $|Ai>$ the possible physical state ai.

In the film *Khartoum,* Laurence Olivier plays the role of Mohammed Ahmed, the Mahdi. As he put it when addressing his Sudanese army, "Oh Beloveds! I am the Mahdi, I am the expected one." Similarly, the ai's are called *expectation values.* They are what we expect to receive as a result of our operations. We use the lowercase a to mean that a physical event has been realized, i.e., registered in consciousness.

We symbolize this law as follows:

$$<I|A|Ai> = ai <I|Ai>$$

We read this as *I-bra-A*-op-ket-*Ai* equals *ai-I*-bra-ket-*Ai.*

This is how the unconscious mind creates the realization of things moving. A-op sandwiches in between I-bra and ket-Ai. It operates only on what it knows. We have an ideal, a psychic sense of moving things. This psychic sense is symbolized by the A-op operator. But how do we tell fast from slow? Fast and slow are relative states of motion that our speed-sense operator $|A|$ distinguishes. When $|A|$ operates on ket-A50, it "determines" the speed of the car to be 50 mph. When $|A|$ operates on ket-A10, it "determines" the speed of the car to be 10 mph.

The meaning of the law is: If I determine (sense or measure using the operator A-op) the state of motion when the state is known to be ket-Ai, will I create the "expected one," ai, in the state I (I-bra)? Can I create a reality by my operations, which in this case is my sense of motion?

The answer to the question depends on I-bra. If I-bra contains state Ai or allows it to exist, the answer is probably yes. In other words, that reality is possible if I allow it. Whether I allow it or not depends on the extent that I-bracket-Ai is possible (i.e., If Ai, then I?).

In quantum mechanics the ket-Ai's are called *proper* states or, in German (they first appeared in Germany), *eigen-states.* They are the physical states of the psychic ideal speed sense. Each proper state is proper and ideal because it is as finely resolved as possible. No proper states contain other like proper states. No proper states are alike unless they are identical. If we know that the car is in a proper state $|A50>$, it cannot be in the proper state $|A49>$ or in any other speed proper state. That means truly ideal. It also means $<Ai|Aj>$, Ai-bracket-Aj, is zero unless i and j stand for the same number.

This property of proper states leads to a corollary of Law III.

$$<Ai|Aj> = 0, \text{ if } i \text{ does not equal } j; \text{ otherwise}$$
$$1, \text{ if } i \text{ does equal } j$$

The meaning of this is again a question. If Aj, then Ai? But this time the answer is no! If the car is moving at 50 mph, it cannot be moving at 49 mph. This says that once a reality is created, once a proper state is realized, reality has occurred. That realization is the appearance of a value, ai. When we "sense" that the car is moving at 10 mph, we "know" ideally that it is moving at 10 mph. (Actually we could be in error, but more about this later.) If the state of a system is proper with respect to a sense operator, the experience will be a "true" experience of that sense. But what if the state is not proper?

Here Law II combines with Law III. A-op only recognizes ket-Ai's. It projects out of ket-I, $|I>$, the possible physical states, $|Ai>$, and produces the physical state ai. But which speed is observed? How fast is the car going? How well can we determine this? From the ket-I, an infinite range of possible ket-Ai's is projected.

To see how this works we shall go through it slowly. We are looking at the question $<I|A|I>$. If I sense the motion of an automobile (using my A-op) will I realize within my I state any expectation value? Can I experience motion? First we will look at $|A|I>$. Then, after we see how motion is projected, we will complete the process with the return to I-bra giving the total double flow of creation $<I|A|I>$, I-bra–A-op–ket-I. So step 1 works like this (using Law IIa):

STEP 1.

$$|A|I> = |A| \text{ SUM (over all } i, \text{ from } i = 1 \text{ to } i = \text{ infinity) of}$$
$$|Ai><Ai|I>$$

Looking at this term by term we have:

$$|A|I> = |A|A1><A1|I> + |A|A2><A2|I>$$
$$+ |A|A3><A3|I> + |A|A4><A4|I>$$
$$+ |A|A5><A5|I> + |A|A6><A6|I> + \ldots$$

A-op operates on all of I's projections onto possible car movements. Following Law III, for each term A-op sees something it recognizes. It then operates on what it knows, the ket-Ai's. This produces another infinite sum that looks like this:

STEP 2.

SUM (over all i, from $i = 1$ to $i = $ infinity) of $ai|Ai> <Ai|I>$

Looking at this term by term we have:

$$|A|I> = a1|A1><A1|I> + a2|A2><A2|I>$$
$$+ a3|A3><A3|I> + a4|A4><A4|I>$$
$$+ a5|A5><A5|I> + a6|A6><A6|I> + \ldots$$

In each term *A*-op has done its work. Each term represents an "expected one." In each term there is an expected value, ai.

Finally reality asserts itself. The bracket surrounds the operation creating $<I|A|I>$, read *I*-bra–*A*-op–ket-*I*, which is realized as a grand sum of all the possible states of motion, written as:

STEP 3.

$<I|A|I>$ = SUM (over all *i,* from $i = 1$ to $i =$ infinity) of
$ai <I|Ai> <Ai|I>$.

The result of all this is consciousness of whatever *A*-op was an operator for. In our example *A*-op was car motion. It could be any physical observable. This consciousness is experienced by the *I* in terms of the expected results, *ai*. The car is moving at 10 mph or it is moving at 50 mph, or 1000 mph, or even faster than light-speed. Each *ai* is the value of the speed expected. How much that speed is realized depends on the *I*-bracket-*Ai*-bracket-*I*. These terms are the double-flowed quamps, the quantum-mechanical probability amplitudes that are the two back-to-back questions reading from right to left: If *I,* then *Ai*? If *Ai,* then *I*? These double-backed quamps are, when multiplied together, probabilities. The greater the probability, the more that *I* has a projection in *Ai* (we could say a "vested" interest), the more likely the value *ai* is realized. In other words, the more likely is the experience "truth."

Automobiles are never perceived to be moving at the speed of light. So if *ai* = the speed of light, $<I|Ai><Ai|I>$ is exceedingly small and more than likely zero. We cannot dismiss it entirely.

Each term $<I|Ai><Ai|I>$ is a complex number multiplied by its complex conjugate (see Chapter 6, pages 156–57). Consequently, each term is a real positive number. It is possible to draw a graph or plot of this number as a function of the *i* in the *Ai.* In other words, for each *Ai* there is a corresponding number, $<I|Ai><Ai|I>$.

Now there are many possible situations, many possible graphic possibilities. These possibilities depend on how much of the *I* state is being realized in the *Ai* state. If all of the $<I|Ai><Ai|I>$'s were identically zero except $<I|A27><A27|I>$, then *a*27 would unquestioningly be the value of the particular observable $|A|$. For example, the car would be observed to be moving at, say, 27 mph. In this case *I* itself would be identical to *A*27 and we would have the case, albeit a somewhat comical case, that "I" am a car moving at 27 mph down the road.

If there is a distribution of the Ai's over I, then Step 3 produces an average or expectation speed of the car. This average value is what one comes to expect as the speed of the car if one operates on the car with the operator $|A|$. For a real car the operation of observing that we all perform when we see or drive already sums up a huge number of auto-speed states. These states are very close together in speed:

50 mph
50.000000000000000000000001 mph
50.000000000000000000000002 mph
50.000000000000000000000003 mph
50.000000000000000000000004 mph
50.000000000000000000000005 mph

Consequently none of our observations even attempts to discern such a small difference in speed. The car may actually be undergoing tiny, tiny quantum jumps in speed each time we do observe its speed, but we never pay any attention to such jumps. Our systems do not possess the required sensitivity to observe such tiny jumps.

This insensitivity to such minute changes makes the universe appear "lawful" and orderly and makes it easier to have agreements about "what the hell is going on around here."

Ego Consciousness As Explained by Quantum Physics

Writing about consciousnesses and egos, I am sure to raise some and bruise a few. After all, how dare a physicist *attempt* to describe anything as hopelessly complex as the human mind? Whether the mind is complex or otherwise, it is possible to describe states of consciousness and even how separate ego states arise. Indeed, ego is a state of consciousness, one that is quite powerful because it is a holographic image of the one mind, the big "I."

Law II described earlier in this chapter stated it. The mind, the one mind, can be composed, coherently summed up, added together holistically, partitioned into projections, and be the sum of all living souls, animal, vegetable, and anything else that lives. Each of these substates of the big I is "conscious." Each big I can be thought of as being part of a bigger I. In turn each little i can be composed of even smaller little i's. It depends on how finely resolved any i-state is made. For example, I am composed of other living entities, other little i's. Each i is conscious. Suppose each i is a cell. Each cellular i is in turn composed of other living things, such as DNA. In turn each DNA molecule is composed of "living" atoms, such as carbon. And each carbon atom is composed of electrons and nuclei, which in turn may be made up of space-time warps or minuscule black holes.

Consciousness exists if at any level the quantum wave function can be popped. That level of consciousness may not be what *you* call consciousness, but if it can observe it is conscious.

In Law II, a separate state was symbolized by *i*. Each little *i* is capable of creation on its own limited scale of existence — in other words, its own limited sense of itself and what it can do or is capable of projecting. The three laws of consciousness can be rewritten with little *i*'s replacing the big *I*'s and some other index letter, *j*, replacing the *i*'s. Thus Law I becomes law i. My mind is mine; "i" am a unit unto myself. This is symbolized in a similar way:

i. $<i|i> = 1$

But how can $<I|I>$ be 1 and $<i|i>$ also be 1? How dare I (oops, i) equate myself with (gulp!) God, the universe, or any other big *I?* It's easy! In physics it is called renormalization. It means setting the scale on which all determination is made. One foot is not the same as one mile. Both are 1. In my limited *i*-sense of the universe, i scale things according to my own scale, my own sense of size.

This ability to renormalize or rescale is a process. I showed how it may occur in the self-reference mathematical scheme in Chapter 3. Each ego state occurs in a similar manner. One defines an operation of reinforced repetitive sequences. Each pattern ultimately stabilizes upon a particular sequence.

In Chapter 3 renormalization occurred when the equation

$$x = 1/(1 + x)$$

was satisfied with $x = .618 \ldots$

This was a one-step sequence repeated until the value of x was realized to as many decimal places as was desired. In the equations

$$\sin(x) = y$$

$$\cos(y) = x$$

where sin and cos are the sine and cosine functions of trigonometry (the sine(x) is the ratio of the opposite side of the angle x to the hypotenuse of a right triangle; the cosine(x) is the ratio of the adjacent side of the angle x to the hypotenuse of the same right triangle), x and y can start off with any values.

With continual repeats this becomes:

$$\sin(x) = y$$

$$\cos(\sin(x)) = x$$

$$\sin(\cos(\sin(x))) = y$$

$$\cos(\sin(\cos(\sin(x)))) = x$$

$$\sin(\cos(\sin(\cos(\sin(x))))) = y$$

$$\cos(\sin(\cos(\sin(\cos(\sin(x)))))) = x$$

$$\sin(\cos(\sin(\cos(\sin(\cos(\sin(x))))))) = y$$

. . .

By running the two-step self-referential sequence many times,

$$\sin(x1) = y2$$

$$\cos(y2) = x3$$

$$\sin(x3) = y4$$

$$\cos(y4) = x5$$

one finds x to be the value .7681692 and the value of y is .6948197 (after twenty or so steps on a TI – 35 calculator in the radian mode). Here there is a convergence on two values.

In a similar manner each ego process yields a stable two-valued I/IT separation. Acts of consciousness renormalize the equation. In fact, acts of consciousness are the stable I/IT renormalizations. They produce or normalize those values which become the "real" world.

This is just like the undifferentiated terms in the preceding section. The ego remains ego-full as the interactions become less and less differentiating or "uncaring." The ego becomes more and more cut off from the world of experience. Through self-generated interaction it splits into an inner schizoid state with each "mind" minding its own business. This occurs naturally enough whenever the mind is uncoupled enough from the external "surprise-and-risk" experiences of the world. The "safe" inner world of the autistic and schizophrenic mind has predictable patterns of ego generation endlessly repeating themselves in a dance of bizarre and beautiful wonderment.

Toward a New Behaviorism

It is my dream that quantum physics will bridge the gap between science and mysticism. As such it must lead thinkers and researchers to a new view of human behavior. B. F. Skinner was not so wrong in attempting to deal with behavior scientifically, but he was the Newton of behaviorists. We now search for the Einstein and the Bohr of human behavior to develop the quantum model of human beings.

Indeterminism and Human Doubt

When each of us doubts the verity of something said or experienced, what are we doing? What is the nature of doubt? If I say that apples have wings, do you doubt me? If you do, on what basis does such doubt arise? If I say that a tree is growing in my backyard, does doubt arise in the reader's mind? Why not? I believe that all statements about reality and even all experiences of reality are fundamentally doubtable. Doubt arises as a naturally occurring phenomenon, based on the fundamental principle of doubt, known as the uncertainty principle of Heisenberg.

We doubt that apples have wings because we have never eaten a flying apple and no one has ever seen one. We base fact on experience and experience on repeatability of experience. "The first time doesn't count" because you can never be sure of the experience. Thus doubt is alleviated when experience is repeatable.

But not every experience is repeatable. In fact, no experience is exactly repeatable. Experience is just what it is when it is where it is. The repeatability of experience is an illusion based on the idea that objects follow laws of motion. A law of motion implies that for any given cause there will follow a predictable effect. That the same cause *always* gives rise to the same effect.

The principle of uncertainty denies all that. It denies the premise. It is never possible to specify the cause in the first place, for what must be specified or known as fact in order to qualify as cause is incapable of being known. Nature insists on her fundamental error. She demands (see Chapter 3, page 100) error in order to create, in order to make new things and evolve. Without this error there is nothing new under the sun.

In Chapter 5, I discussed the problems of Maxwell's demon as he tried to beat nature's error and discern the speeds of fast- and slow-moving particles. I called his discernment ability "Δv." This was the difference between two speeds, $v1$ and $v2$.

$$\Delta v = v2 - v1.$$

The smaller this difference, the greater was the ability of the demon to put the faster particle on one side of the box and the slower particle on the other side of the box. Now, in order to separate one particle from the other, by telling if the particle being examined was faster or slower than average, the particles were allowed to approach a doorway covered by a slide in a partition separating one side from the other. The demon operated the slide as fast as he could, but if the doorway was too large, too many particles would pass through before the demon could cut them off. This would spoil his attempts at speed separation.

The uncertainty principle prevented the demon from making the slide-doorway aperture too small. Here we want to look at this limit and the error it produces.

It works according to a simple formula that says that the product of human discernments of the location of an object in space and the momentum of that object must not be less than Planck's constant, h, which is 6.6×10^{-27} gram-centimeters squared per second. Now, this tiny constant of nature is perhaps understood a little better if we go down in scale to the size of an atom (about 10^{-8} centimeters, called one angstrom unit) and look at its effect on the tiny electron (whose mass is 9.1×10^{-28} grams). Furthermore, the time scale on which this comic theater of errors operates is around one nanosecond (10^{-9} seconds). On this scale Planck's tiny error constant becomes quite large and is

$$h = 73,000,000 \text{ egrams-angstrom squared per nanosecond}$$

where an egram is the mass in grams of one electron. To see how h operates, imagine putting the electron on a line so that we know its location within a stretch 1 angstrom wide. Within one nanosecond that electron will, for no reason other than its confinement to the line, possess a speed of 73 million angstroms per nanosecond. In other words, within that short period of one nanosecond, the electron could be 73 million angstroms toward the left or right of the original stretch. Now 1 angstrom is to 73 million angstroms as 1 inch is to 1152 miles! If the state of California (which is about 1000 miles long) was 73 million angstroms long, an electron put in the middle of San Diego would, in a billionth of a second, "appear" at the tip of Baja or on the Oregon border!

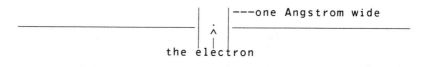

Figure 55

Just confining the electron this much introduces this huge Δv in its speed and therefore a wide spread in its possible locations just one nanosecond later. The speed of light in angstroms per nanosecond, by the way, is 3 billion, only forty-three times as fast as the electron's Δv.

This means that it is impossible to specify the location of the electron to within 1 angstrom of width without "allowing" the electron the considerable Δv in its speed of 73 million angstroms per nanosecond. In terms of atomic-size California, you give it an inch and it takes not just a mile but 1152 miles in the next billionth of a second. To attempt to put the electron on the line, confined to a strip 1 angstrom wide with a specified speed of less than 73 million angstroms per nanosecond, is impossible according to the uncertainty principle. We cannot be certain at the same time of its speed within this considerable speed range and its location within 1 angstrom.

Thus we cannot say that any movement is repeatable because, in order to repeat, movement must cause the moving object to return to the same place with the same speed at a different time. Thus there is a reasonable shred of doubt in every person's case. There are no absolute objective facts at all. With every idea, with every experience, every heroic jump into the unknown, every act of cowardice, each morsel of food digested, each thought of beauty or ugliness, there is doubt.

The trick is, as Lawrence of Arabia, in the film of that name, proclaimed when showing an amazed noncommissioned officer how he put out a burning match with his bare fingers, "in not caring that it hurts." The trick is we do what we do in spite of the doubt that is always there. The trick is in not caring that doubt is there. Doubt is necessary in order to have a universe.

Doubt and Survival

Richard M. Restak, in talking about the mind, takes a psychobiological approach. He concludes that to consider both the brain and the mind as things is to make a logical category mistake. Instead the brain is a thing and the mind should be regarded as a process. Thus for Restak the mind cannot have a physical location any more than *red* or *feeling* can have a physical location. Restak concludes that the world as we experience it is a product of our brains, and only by understanding how our brains work will we reach true insights regarding human motivations.

Restak and Charles Hampden-Turner write about the work of brain researcher Dr. Paul MacLean, who has done much to advance our knowledge about the "big grey walnut lying above our nostrils." MacLean believes that we don't simply have one brain but three: the reptilian; the old, or paleomammalian; and the neomammalian, or neocortex. The structure of these older brains resembles the corresponding brains of reptiles and the early mammals. Our brains contain the history of our evolution in much the same manner as the Grand Canyon contains the geological history of the earth, in layers. The lowest, or reptilian, brain consists of the brain stem, the midbrain, the basal ganglia, and much of the hypothalamus and reticular activating system (RAS). Tests have shown that it contains large amounts of dopamine, a neural transmitter. This brain is considered to be a slave to the higher, old mammalian brain encompassing it.

The old mammalian brain consists of the limbic system and comprises two nearly concentric rings, one in each hemisphere. This brain is responsible for reward and punishment, emotions, and controls of the autonomic functioning of the body.

The new mammalian brain, or thinking cap, represents the latest evolution. Mostly a surface area, it is a convoluted mass of gray matter that if spread out would equal the size of a small rug.

Hampden-Turner poetically calls MacLean's research, which is based on the work of James W. Papez, "Lying Down with a Horse and a Crocodile." This phrase refers to MacLean's statement, "We might imagine that when a psychiatrist bids a patient to lie down on the couch, he is asking him to stretch out alongside a horse and a crocodile."

Human behavior is thus not a simple cause-and-effect, stimulus-response system. Three separate computer networks, corresponding to the three brains inside of our heads, are at work. Often experience results in conflict among these three. What we sense as needed for survival (our reptilian mode) may be emotionally disgusting (don't eat that; it's been hanging there for days). Our intellectual thinking caps attempt to mediate between these conflicts. As a result doubt and uncertainty arise.

But hold on. Maybe we've got this all reversed. Perhaps our brains evolved the way they did because of the universal requirement of the uncertainty principle. Our "geological" brain layers may be required in order to allow doubt and conflict, which are sensed forms of the uncertainty principle, the "space" needed for correlation that is necessary for evolved brain functioning and intelligence. Like the whole universe, our brains are quantum mechanical, not classical mechanical.

Our human complementarity follows from the quantum complementarity. Similarly, the brain functions in a hierarchy that evolves in space through correlations arising out of this fundamental quantum-physical complementarity. Horizontally, the brain operates complementarily. Vertically, it operates correlationally.

THE QUANTUM BRAIN

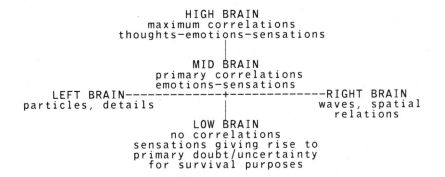

Figure 56

The lowest, or reptile, in us fails to correlate data. Data are stored in simple patterns of stimulus/response. The reptile is a machine in this sense. Its conflicts and doubts arise fundamentally from indeterminism, not from thought or emotions. "Fight or fornicate" is determined by a simple analysis of a series of events already prewired from previous ancestors, who managed to survive by knowing when to fight and when to fornicate. Given enough clues, one behavior is probabilistically indicated over the other. The odds are arrived at by ritualistic repetitive behavior. The animal does its dance. If it repeats the dance, the message is clear: "Let's get it on, baby." If it breaks the pattern or shows a different one, it's time to fight or flee.

The midbrain experiences spatial area while the low brain experiences online one-dimensional sequences only. Sensations are correlated, resulting in experiences without thought but which are more than just sensations. These experiences are the emotions, or "felt" feelings, as differentiated from Jungian feelings. These feelings arise out of quantum correlations from different parts of the midbrain. The ring structure of the limbic system orchestrates these correlations and involves both halves of the brain simultaneously.

The high brain, or neocortex, introduces one other correlation experience. Sensations and emotions are correlated and symbolically stored. This new dimension of symbolic storage arises from, and integrates over, the lower brains. Conception is the symbol correlated out of the experiences of sensations and emotions. In this manner consciousness "learns" about the world through its symbolic map that reenacts the emotion-sensation correlative experience. One picture is worth a thousand words. One word is worth a thousand pictures.

Correlations of conceptions (or superconceptions) result in creativity and human culture.

Death and Reincarnation

What happens to our minds when we die? Our bodies obviously undergo radical transformations. I remember being mystified by a question about the mystery of the life/death transition put to me by the Hawaiian kahuna, Abraham Kawai'i. He told me that there was an ancient biblical riddle that, once answered, was the solution to the age-old mystery of survival. He asked, "From what do the bees in the desert arise?"

I responded by saying that I didn't think that there were bees in the desert. How could there be? Don't bees need plants and flowers? I know that there are desert plants but, if I were a bee, it would seem logical to me to spend my time and build my hive in a more habitable environment.

Abraham told me the answer. Life feeds on death. As a student studying the ancient code of the kahunas he actually witnessed bees arising out of the dead carcass of a lion. When animals or plants die, life in the form of other animals or plants takes over and lives off the carcass. Lion carcasses give rise to bees. Death gives rise to new forms of life.

So much for the body. Now, how about the mind? Does the mind survive death? Although there is much written about the processes of survival, it appears that no one knows for sure. Here I shall attempt to speculate and, as always, base my speculations on the conceptual framework of quantum physics.

First of all, the mind is not a physical thing. It is an idea or process. What we usually mean when we say "my mind" is not at all clear. For example, the processes we call mind seem to occur both inside and outside the confines of the brain cavity. Our ability to locate experiences at their sources (sound of rain on the roof is not just heard in our ears but is projected out there to the drops spattering on the surface) indicates that we can reconstruct literally anywhere the experiences in our minds. The only reason we don't normally think of our minds as outside our bodies is that we feel with our skins, taste with our mouths, and smell with our noses — use organs that are primarily senses confined to "surface effects," while hearing and seeing are three-dimensional or holographic effects. This limiting of three of our senses outweighs the seeing and hearing senses and convinces us that we live underneath our skins.

If we were able to taste, smell, and feel a tree 50 feet away across the yard, it would never occur to us that our minds were inside our bodies.

From the viewpoint of quantum physics espoused in this book, mind is the process of qwiff pops. These pops occur simultaneously both inside our heads and outside at the sensed experience. For this to occur the quantum wave function must flow from A (inside our heads) to B (outside, at the sensed experience) and back again from B to A in the time-reversed sense in order that the $A \rightarrow B \rightarrow A$ sequence is experienced. This wave flow is associated with a frequency or a repetitious pattern in time. This frequency is associated with energy according to the Planck-Einstein discovery:

$$E = hf$$

where E is the energy capable of being experienced in the transition from A to B, h is Planck's constant (see my earlier book), and f is the frequency of the quantum wave function.

Energy comes in many different guises. Before Einstein's recognition:

$$E = mc^2$$

there was chemical energy, heat energy, light energy, electromagnetic energy, mechanical energy, kinetic energy, and potential energy, to name a few of the forms. That these energies were also in some sense equivalent is what gives energy its usefulness in the first place. As I mentioned in Chapter 1, mechanical energy such as in a rolling wheel can be transformed into heat energy by rubbing it on a rough surface or into electrical energy by having it turn a bar magnet in a coil of copper wire.

With Einstein's discovery, all energy was recognized as *matter!* Matter was the driving, or hidden, energy form to be transformed whenever energy was moved or transported from one place to another. Thus when a wheel rolls, it has more mass than when it is at rest. The excess mass is converted to heat when it rubs against a rough surface. When we write down the energy of the rolling wheel, we say

$$E(\text{rolling wheel}) = E(\text{wheel at rest}) + E(\text{motion})$$
$$= E(\text{wheel at rest}) + E(\text{hot surface})$$

or

$$M(\text{rolling wheel}) = M(\text{wheel at rest}) + M(\text{motion})$$
$$= M(\text{wheel at rest}) + M(\text{hot surface})$$

The heated surface is not only hotter, it is heavier as a result of acquiring energy/mass from the rolling wheel. When an atom emits a photon of light, it too loses mass. The photon takes up the mass given off. When that photon hits another atom and is absorbed, that atom increases its mass by just the amount given up by the previous atom. It is the mass "ball" or dance that compromises life in all of its energy processes. The power of Einstein's recognition lay in the possibility of transforming matter at rest into heat or light.

From the Einstein/Planck $E = hf$ formula, energy is also frequency. This means that at the quantum-wave-function, or qwiff, level of reality, energy must also be capable of undergoing transformation. I believe that what we call the spirit of life is capable of manifesting as both matter and life energy. Thus the frequency of the qwiff corresponding to this life spirit is also convertible or capable of undergoing transformation. I believe that life's processes and death's processes differ by a qwiffian frequency. This frequency can be associated with the process of "paying attention." It takes energy to pay attention. Life is different from death only as a degree of that attention. We could call this "paying attention" consciousness. Thus we could write for the frequency of the qwiff:

$$f(\text{living body}) = f(\text{dead body}) + f(\text{attention processes})$$

But this is really a gross simplification. A better way to write this is:

$$f(\text{life force}) = f(\text{physical processes}) + f(\text{mind processes})$$

Before there is a physical body the frequency of the life force is all mind. As matter appears, the life-force frequency is split into two parts just as the rolling wheel's mass is split into two parts, a rest mass and a kinetic mass. With growth and maturation the mind frequency decreases and the physical frequency increases. With the approach to death there is a release and a quickening of the "pace" associated with the return of the mind frequency becoming all life force. The dance of transformation from $f(\text{physical processes})$ to $f(\text{mind processes})$ and back again alters the qwiffian waves.

In Figure 57 we see the general movement of these two frequencies. The trend or cycle is as shown. However, if we were to look at these frequencies in a very tiny time interval, they wouldn't appear so smooth. The smaller the time interval, the harder it is to tell which is which. The universal processes are governed by the uncertainty principle, in this case with energy and time as the variables. Since $E = hf$, this means that the frequencies are changing (Figure 58).

On the tiny time scale of quantum electrodynamic effects, much higher frequencies are spontaneously generated in processes that are quite similar to

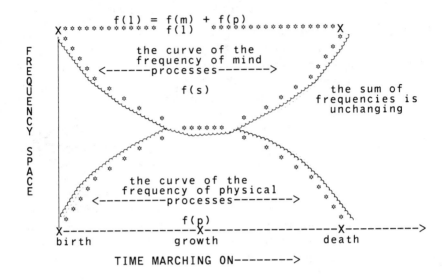

Figure 57

life and death. These are called annihilation and creation processes (see Chapter 6). When attention is paid to processes on such a tiny time scale, life as we know it completely vanishes. It appears utterly chaotic. As the time scale is stretched or lengthened, order begins to appear through long-range correlations brought about by stable interactions between these chaotic events. The stability of these interactions, such as the atomic stability of all known physical elements, is a compromise between the chaotic forces of qwiff popping and the organizational long-range forces of interaction like the electromagnetic force.

The game of attention is the freeing of matter from its "frozen" frequency condition (i.e., $mc^2 = E = hf$). In other words, the qwiff high frequency is "stuck in a feedback loop" of enormous repetition, seemingly endlessly reproducing itself as inertial mass. Recent research, however, indicates that matter may not be completely stable. The proton is now predicted to decay eventually through *electroweak* forces.

All matter is ultimately unstable. (It has not yet been shown that the electron will decay by any known processes. I intuitively predict that within five years the electron decay will be predicted.) Eventually consciousness upsets the stable patterns of matter and frees it, turning it back into light. Meanwhile, back on the ranch of everyday occurrences, this dance is much, much tamer. The human life cycle is the dance of the higher mind frequency

212

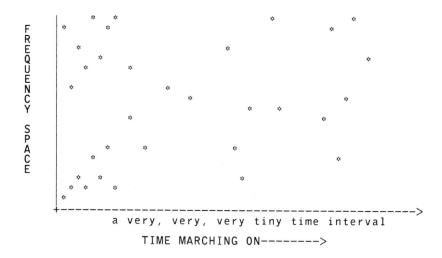

Figure 58

with the lower material frequency. This higher mind frequency always has the capability of manifesting as matter. The electrochemical dance of matter in the brain and nervous system is obtained through the interference patterns between the mind and the physical frequencies.

Again, I am providing a rough sketch of a quite complex process. The mind-matter interface is an interference pattern, with mind tending to condense into matter and then escape again, and matter tending to turn into mind and then condense again into its frozen or trapped qwiff state. Death comes when that dance ends. It ends according to some plan or agreement that ultimately relates to the decay of all matter into light.

What we call reincarnation is attributable to two possible explanations. The first is that time is an illusion. What we call matter and its movement in time is an interference pattern between an infinite number of parallel universes. Consequently, we are simultaneously alive in all of these universes. The past "you" is in an interference pattern with the present "you." The past "you" experiences "you" as a dream or thought. In a similar manner you "remember" or "imagine" your past life. I'll return to this in Part Four.

The second explanation of reincarnation is genetic. Our genes contain molecular qwiff patterns. These patterns were set up according to experiences of our ancestors. Hence there are tendencies for repetition of events that were influential in the past. These events may not have actually happened to our bloodline forefathers and foremothers, but these bloodline parents were never-

213

theless influenced by them in some manner of significance. Your great-great somebody may have slept with Cleopatra or Genghis Kahn. That imprint lives on as a genetic influence today. If you trace back only thirty-two generations, you will count over four billion ancestors!

Finally I want to propose an answer to the age-old mystery: What happens to the mind when you die? If we equate the mind with that sense of ours that is prethought and preexperience, then it continues without end. This is called many names: the Transcendental Self, The Self, God, The Original "I," The Spirit, the Soul, Aleph, Elohim, the Transcendental Personality. This mind, which I call pure qwiffness, is a qwiff without matter. It is the life force that is beyond life/death. It is a qwiff without a pop. It is potential able-to-be-ness. Looking at Figure 57 once again, think of those kissing frequency curves as sinusoidal rhythms. They will repeat from lifetime to lifetime. Each passing generation awakens to the spiritual descent into matter. It is as if an artist (God) sees the world as canvas. With each stroke of the brush a little piece of God consciousness is left behind, momentarily trapped as part of the living genetic code enlivened through quantum correlations. This slight code alteration dies out unless it is passed on through sexual reproduction or education. With each evolved transformation the new generation is able to play a more enlightened "game/war with time." The game never ends.

Personality, Observation, and Quantum Physics

Consider the following thought: There is no "out there" out there. All that is is continually being created out of nothing. You and the objective "out there" are part of that ongoing process of continuing creation.

The "out there" depends on us for its existence; and we in turn depend on it. Our knowing that it exists is directly coupled to our sensing that it is there. This sensing depends on our interaction with it. That interaction comes to be known. That process cannot be separated from the act of observation. Our thoughts, which attempt to organize our experience, classify it and compare it with our memories of previous experiences, come later. We often tend to confuse those later experiences with the actual act of observation. Our knowing of the chair and our experience of the chair are really one and the same thing. Ontology and epistemology are not independent.

If the physical world, then, is dependent on us, on our acts of observation, and in turn these acts are what we call real, doesn't it follow that what we call personality is also dependent on us? Doesn't it follow that the way a person appears to us, his or her persona, depends on our acts of observation of that person? Isn't it feasible that what we call a person's personality depends on that person's interaction with the perceiver of that personality? Since we ourselves are all different (a debatable issue, I agree), wouldn't it follow that a person's personality is a measure of that person's interaction with the perceiver?

If the perception of people is similar to the acts of observation we all perform as we observe the physical world, then people's personalities cannot be innate. They too must follow the laws of quantum physics. They too must be continually created. This leads to a whole new study of personality. This study begins with the assumption that personality is not an attribute but a process. Since it is a process, it is creatable, changeable, and has no existence independent of the acts of its observation. Surely this must have an impact on psychology. Perhaps we will renew our understanding of the term *personality*. It is an ever-changing mask. It is the creation of persona, which means mask. There is no personality. There is no "self." The mask is all that is, and that is process.

The Universality of Suffering

The Buddhists claim that suffering is universal. It arises as a necessary outcome of the chain of interdependent causes traceable ultimately to one cause: the false identity or ego we all seem to possess. Thus suffering and the ego are the same thing.

What is this suffering? How do we experience it in our Western haven of plenitude? Recently I picked up a copy of *The Plain Truth* and read an article entitled "Why Must Man Suffer?" The writer, Herbert Armstrong, is well known as the editor-in-chief of this monthly religious magazine with a circulation of over six million. I was surprised to see how close Armstrong's Christian interpretation of suffering was to the Buddhist's. According to both, the answer can be described in one word, *choice*. The choice to identify with a particular point of view that denies or cuts off another creates a surface of separation, dividing in our own brains the flow of consciousness, or qwiff flow, into separate pockets. The result of this is human suffering. When we feel a flash of intuition or a knowing sense of truth, we are relieved and experience surrender to something bigger than our pitiful selves. This event is the sudden release of the qwiff from a trapped condition. Suffering is no more then the enforced pattern of repetition or habitual behavior produced by our illusionary senses of inappropriate boundaries. These boundaries are created by the symbolic forms, usually words, we choose to describe a situation. When the word is spoken again or referred to as a descriptor of a novel physical situation, suffering occurs because the new experience doesn't match the invoked experience brought to our minds.

Typical examples of this kind of suffering can be heard at the dinner table. Dad comes home tired. The boss complained because Dad didn't perform his job up to expectations. Mom has a headache. Daughter is going out with a black (or, if this is a black family, she is dating a white). Son is smoking pot. The neighbor's dog barks and whines continuously. The crime rate is up. The Marines in country X are being killed. The population is increasing. Joe is gay. Ethel is in a mental institute. God is either dead or weak and ineffectual.

Inflation is wiping out our savings. Need I say more? Aren't your complaints somewhere in the above list?

Each and every situation arises because each and every situation is entirely new. We all attempt to seek the past for answers to the present and thus fail to understand that these situations are inevitable consequences of the choices we make and the chances we take with life. We cannot help or stop this process of risk and karma buildup that results from these choices. These arise as inevitable consequences because the Heisenberg principle of uncertainty and the Bohr principle of complementarity operate in our daily lives.

Suffering always arises because of the inappropriate separations we feel from these problems. We each fail to see the other side of the "problem." It is this failure that makes it a problem. The other side of the problem is nothing more than the cognition that if anyone is suffering, *you* are suffering. If one blade of grass is crushed, something in you is crushed as well. To recognize this is to understand the inevitability of suffering. To relieve the suffering is to free yourself from any repetitive patterns of self-identification that you feel. This is usually difficult because it is a humbling and embarrassing feeling.

If Joe is gay and Tom shoots heroin, we are gay and we are heroin addicts. This follows from the inevitable senses we have. To know of the existence of suffering is to feel that suffering in yourself.

We learn of suffering through electromagnetic interactions usually brought to us by the communications media. The 11:00 P.M. news even provides the opportunity for us to "sleep" on the suffering of the world. News is always bad. (Good news is not news to the media.) We *need* to hear bad news. No one watches the news to find out how well others are doing. We need to be reminded of how well off we are by looking at the suffering of others. To relieve us of this reminder of the "real" world we watch commercials. Here we find out that we can relieve our suffering, our sagging waists and breasts, bad-smelling breaths and armpits, and vaginal and athletic itches by watching others like ourselves squeezing toilet paper or by imitating fantasy images of impossibly beautiful people bustling out of pools with soda pop or perfume in their hands.

But, sad to say, we still smell bad in the morning, and our waistlines continue to grow as we overeat in frustration and starvation, failing to realize that as long as hunger exists we feel it.

To relieve such suffering is to recognize that each of us has created it. There is hope only if we each choose to hope. There is none if we each choose that there shall be no hope. We are the creators of *this* universe. We shape its raw materials into our fantasies, which we call realities. We have reached this stage of understanding in the most peculiar way: the Western way of science and industrialization. We have learned to manipulate the physical and have discovered the spiritual.

The next time you find yourself suffering, try this. Become what you hate for a moment. Do this in your thoughts and discuss this with your friends and families. Turning the other cheek or loving your enemy is no act of foolish charity. It is a real solution to the problem. By accepting our foibles as *our* own we will gain that quantum correlation that frees our trapped qwiffs. For a brief moment we will relieve our suffering.

Toward a Theory of Bliss

In *Taking the Quantum Leap* I stated the idea that quantum physics and human consciousness were intimately related. The basis for this is the role of the observer in the act of observation. He/she is not passive. Instead he/she plays a unique role, which depends on what he/she believes is "out there." (If you look for waves, you find waves; if you look for particles, you find particles.)

Thus humans play a far greater role in their own destinies than they may have originally thought. Indeed, we seem to be ruled by our projections of what is and what isn't. We live by our wits by projecting our abstractions into the physical world. An example is the concept of the state.

The state is pure abstraction, yet we all believe in it and in its power to rule us. We create physical buildings and state capitals that embody the state. We feel imprisoned by it. Thus our abstraction takes on a superlife of its own. Doesn't it follow that we can create through our acts of observation the *bliss state?*

Not only does it appear to be possible, it is happening and has happened. Spiritual teachers (albeit some phony, but most quite genuine) abound today. If we believe them, the bliss state exists. It is a teachable concept open for those who wish to learn it. It can be learned in the same way that any seemingly complex skill can be learned. (Most spiritual teachers would probably say that one needs to get rid of the intellect in order to do this. Nevertheless, most of us need guidance on how to do this.)

Even if we don't believe these spiritual teachers, we need only look at quantum physics to realize that the physical world cannot and does not and will not exist without us. We are necessary for it to be!

Paraphrasing the words of visionary physicist John A. Wheeler, there is absolutely — and I mean totally and undeniably — no elementarily known phenomenon of nature until it is an observed phenomenon. For example, all atomic experience is of this nature. When we are dealing with macroscopic observation we are dealing with extremely large numbers of fine microscopic observations. Our nervous systems are continually bombarded with these micro-observations; and these micro-observations are guided by choices that ultimately we have made through our evolutionary ability to project abstract thought into living experience. Thus the time is now for all humans to learn to be blissful.

What is the bliss state and how do I reach it? First one realizes that no "tangible" substance gives bliss (but it helps — at first). Bliss is a state of consciousness. I believe that it can be reached by individual effort (although like any other skill it eventually will be effortless). In this state there is a divine "ignorance." Actually *ignorance* is not the correct word. A better choice is *unknowingness.* There is an awareness of unknowingness. It is akin to the feeling that an artist has when he/she is creating. There is a faith in the ability to create, even though it hasn't manifested as yet. There is a realization that "I" can create this. There is the identification of the "I" with the process of creating. There is the realization and eventual discovery that all is constantly and ever-changingly new.

This unknowingness is fundamentally quantum in nature. It depends on the underlying reality of the quantum wave function (qwiff) — the wave of all possibilities. Faith in its existence of unknowingness starts the process of bliss. In bliss there are no victims, no manipulations, no guilt, and no realizations but the realization of no realizations. Realizations manifest and vanish like thoughts.

Paramahansa Yogananda has written a remarkable little book about the bliss state of consciousness, called *The Science of Religion.* Yogananda indeed believes that bliss can be learned and sets out to explain how. He points out that bliss can be neither purely self-serving nor purely other-serving.

Even the most altruistic motive and sincerest intention of advancing the good of humanity ... have sprung from the basic urge for a chastened personal happiness, approaching Bliss. But it is not the happiness of a narrow self-seeker.

The motives of all people are bliss seeking and pain avoiding. Thus religion arises as the means by which the state of bliss is to arise. If we take as a tautology that the removal of suffering and the acquiring of bliss is religion, then all humans are religious. In this sense of religion the "new physics" is a religion, or will become one, because it appeals to that sense of enlightenment and mystery that we intuitively sense as truth. It offers a solid basis for understanding what is meant by the spiritual or mind element.

The universality of all religions is the recognition of the connections, or correlations, we have to all living things. This connection is like a chain. Whoever pulls it creates a feeling for whoever is attached to it. We cannot help but feel the chains of attachment created by our physical interactions. These interactions tend to correlate us, providing us with a sense of unity. It is our continual repetitious patterns of separate egos that act to destroy and break the connections. This is the complementarity of the human mind in action.

Yogananda states that our lack of bliss, or our suffering, is brought on by the process of identification with the transitory body and the restless mind. In my quantum view the qwiff is trapped into continual awareness of the boring details of mind and body in endless pursuit of happiness. It is continually being bounced around by sequences of operator/observables, which produce disruptions of bliss states. A sequence of self-referential or biological feedback operations alters the state, tending to produce distributions of events that are repeatable and directly sensed as nonthreatening and survival enhancing. These operations are inherited from our reptilian and mammalian pasts (or, as I will point out in Part Four, from our recreation of the mammal and reptile in our prethought patterns). They appear to us as powerful urges that at our present state of evolution are no longer needed for either survival or pain avoidance. Indeed, today they act in just the opposite way, threatening us and producing the means of our extinction.

The way out is through the recognition provided by the insights of quantum physics. It is vital to realize that matter is not king of the universe but its slave. As such it rebels to escape. By realizing that all matter is time-bound, but qwiffs are bound by neither matter, energy, space, nor time, we have a choice of identification. It is no longer necessary to identify with your fear and your survival. Identify with your spirit, your qwiffness, and we all become free. You are holding us back. You are the liberator.

Fear, Death, and Narcissism

Although it may not be obvious, Freud was extending beyond mechanism in attempting to explain the human psyche. He was not only the Newton of psychology, he was also the Bohr. He was trapped by sexual energy concepts prevalent in his time. To Freud the libido was energy. I would like to offer the idea that the libido is trapped qwiffness and not energy at all. Freud was mistaken in his male chauvinism. To the intellectual spirit of his time the feminine was thought to be subhuman. Knowing or believing this meant that all humans were subhuman to a degree, trapped in narcissistic self-consciousness.

Narcissism is the drive for self-preservation. It is the self-referential process that defines the "I/IT" separation. It continues as habitual sequences of renormalization. Whenever an experience tends to alter a self-reference feedback loop and thus broaden the "I" sense, pain is felt. Paradoxically, when we avoid this pain, the pain becomes acute and suffering results.

For example, in thinking about giving a lecture or providing some service I think of my survival: How much will I receive? How long will I have to talk? Will there be hostile questions? This is narcissistic. If I recognize the service that my talk renders to others, if I really experience *their benefit,* I also benefit but in an entirely different manner. This benefit is complementary to the self-reference experience of survival. One must operate in a complementary mode to receive it. By only being concerned with narcissism, we lose this life-preserving benefit freely given by others. Learning this can be the way out of inappropriate cathexis. It will result in the creation of released libido.

If the pulling in or, as Da Free John puts it, "the ego is self-contraction," is taken to be the life drive, it also results paradoxically in death. Absolute narcissism is death. Most people are not aware of this. Take pleasure, for example. Although we all need pleasure in our lives, there is definitely a need to stop feeling pleasure. If that need is not met, the resulting hedonism leads to abuse of substances. Substance abuse is total narcissism. It creates death because it cuts off life from the outside world. Total pleasure and comfort is death.

Absolute narcissism is total death. It is the complete withdrawal of libido into maximum self-consideration. If the individual is the expression of 100 percent narcissism, the individual dies. This would be the total contraction of all experience into oneself. Birth is the opposite experience. It is the total explosion of oneself into the universe. Death is a contraction into oneself; birth is the expansion out of oneself. In each person these two drives exist unconsciously — the life-death grid. Without this grid one does nothing. There is no motivation. Birth and death must be in balance, like the law of physics — often called a superlaw — that electrical charge is conserved. For every plus charge there must be a minus charge. Thus no living individual can be totally narcissistic.

What about death? There are two kinds of death:

- The passing of the physical body and
- Living in a state of consciousness of thought without movement

In each state there would be no narcissism. In this view there would be no outside object existing as the object of thought. Thought would be the object of thought. "You" would not identify with yourself but with the whole world. Everything becomes you. This would be a death-of-the-ego experience. Your ego is now the whole world.

The ego, taken to be a cathexis of libido or pattern of repetition within the unconscious, is a contraction — a formation of "mass" in the unconscious. It is the psychic equivalent of the m in Einstein's equation $E = mc^2$. This ego does not act. It can only react to unconscious influences. It takes on emotional charge and is buffeted by the polarities of the unconscious like a charged mass in an electric field.

How do we experience no movement in time-thought? When you are aware of the benefit to others of the result of your action — the experience that others are benefited, uplifted, or downgraded, etc., as a result of your actions — that sense or feeling you have is the opposite of narcissism. You experience this as the joy of giving, for example. If, on the other hand, you do this with the expectation of reward, it is narcissistic. This sense of "benefit to other sentient beings" is felt by you only when it is not desired.

The feeling of fear is always coupled with narcissism. We experience this as limited egos. This contraction is universal. It is the collapse of the wave function. Through this quantum-mechanical wave-function collapse an object-cathexis appears.

In brief, narcissism results as a process of self-identification. It arises as the need to separate the "I" from the "IT." By reevaluating or altering the way we do this — as for example by including external interactions (others) and identifying with those external processes that are the most uncomfortable to us (and thus changing the probabilities or pattern of qwiff pops) — we will become blissful. Religionists may call this God. I call it quantum physics.

Complex Numbers and Probability Amplitudes

In the example of the ringing telephone I pointed out that the sequence corresponding to $<B|A>$ (if the phone rings, A, do I hear it, B?) will become realized if the time-reversed sequence $<A|B>$ (if I hear the phone ring, B, has it rung, A?) is generated by the conscious mind. As fantastic as this sounds, there is a further justification for this idea. This arises in the mathematical form of the bra-kets.

In order to represent the physics of things, quantum physics has had to utilize the mathematics of complex numbers. A complex number is composed of two numbers in a precise relationship. These two numbers provide two kinds of information, amplitude and phase, which are needed to represent the movement of quantum wave functions.

The bra-ket is just such a complex number. But how do we make any sense of or attach any meaning to this number? Quantum mechanics says that to convert the information contained in $<B|A>$ to something useful we must perform a multiplication. We must multiply $<B|A>$ by a number, called the *complex conjugate* of the bra-ket and written $<B|A>^*$. The asterisk is intentional.

These bra-ket amplitudes, or quamps, are added and multiplied following the rules of complex numbers in mathematics. Each complex number amplitude consists of two numbers, a and b, which are not to be confused with the events A and B. The first number, a, is called the *real* and the second number, b, is called the *imaginary*. Together they are written as a pair of numbers in a parenthetical expression (a,b) that satisfies a not-so-obvious arithmetic scheme.

When two complex numbers, (a,b) and (c,d), are added, the result is $(a + c, b + d)$, as we might expect.

$$(a,b) + (c,d) = (a + c, b + d)$$

But when (a,b) multiplies (c,d), the result is a not so obvious $(ac - db, ad + bc)$.

$$(a,b) \times (c,d) = (ac - bd, ad + bc)$$

Thus, for example, $(1,2)$ times $(3,4)$ gives $(-5,10)$.

$$(1,2) \times (3,4) = (1 \times 3 - 2 \times 4, 1 \times 4 + 2 \times 3) = (3 - 8, 4 + 6)$$
$$= (-5,10)$$

The number $(a,b)^*$ is defined to be $(a, -b)$.

221

So if we multiply (a,b) by $(a,b)^*$, we get as a result $(aa + bb, 0)$.

$$(a,b) \times (a,b)^*, = (a,b) \times (a,-b) = (aa + bb, a(-b) + ba)$$
$$= (aa + bb, 0)$$

The first number, the real part of the product, is the sum of the squares of the two numbers a and b. The second number, the imaginary part of the product, is 0. Thus whenever a complex number is multiplied by its complex conjugate, a real and positive number is created.

The fact that the bra-ket $<B|A>$ is a complex number means that it has both a real and an imaginary part. Physics only deals with real things; yet in order to understand real things we must use this complex type of numbers. Thus, this multiplication, $<B|A>^*<B|A>$, although quite common in a mathematics course on complex variables, is strange in classical physics. The manifestation of reality is obtained from this since complex-conjugate multiplication will always produce a number that is real and positive.

Having a scheme that always produces real and positive numbers is heartening to physics. The next step is interpreting the real positive number you get. In quantum physics, $<B|A>^*<B|A>$ is interpreted as the *probability* that the sequence of events actually occurred. The larger this number is in comparison with other probabilities, the greater the likelihood of the event sequence taking place.

There is a well-known special rule in quantum physics . It says that

$$<A|B> = <B|A>^*$$

If this funny little rule were not true, according to quantum mechanics, probability rules as we understand them would no longer be valid. As a consequence of this invalidation the rules of the physical universe, such as the law of energy and matter conservation, would not be valid. Indeed, if this quantum mechanics rule suddenly vanished, all of physics would go up in a puff of smoke and so would we.

This quantum rule says that the probability amplitude for event A to occur given that event B has actually occurred is the *same* as the probability amplitude for event B to occur given that event A has actually occurred.

This would mean that it is just as likely that I hear my phone after it rings as having it ring after I hear it. How could this be?

Well, this is not quite it. There is this business of the asterisk. It says instead that the probability amplitude for event A to occur given that event B has actually occurred is the same as the complex-conjugate probability amplitude for event B to occur given that event A has actually occurred. Let us use our imaginations for a moment and remember that A is the telephone ringing and B is our hearing the phone ring. It's certainly the usual situation that if the phone rings, you hear it — but not always.

There is always a slight chance that you won't hear your phone ring. Thus $<B|A>$ means the phone rings and *next* I most likely hear it.

But what could $<B|A>^*$ mean? Suppose I wrote the complex number for $<B|A>$, and it was (a,b). The complex number for $<B|A>^*$ would then be $(a,-b)$.

$$<B|A> = (a,b)$$

$$<B|A>^* = <A|B> = (a,-b)$$

They are somewhat alike. If we think of that second imaginary number, b, as a mirror, then these two numbers are mirror images of each other, like reversing right and left hands. (Actually a real mirror does not reverse right and left; it reverses front and back. Thus if you were 5 feet in front of the mirror $(+5)$, you would appear to yourself to be 5 feet in back of the mirror (-5). Your right side is still on the right, but because the front and back are reversed it "appears" to be on the left.)

Thus $<B|A>^*$ can be thought of as the time-mirrored image $B \rightarrow A$ of the possible sequence $A \rightarrow B$. Indeed, since $<B|A>$ means how probable is B given that A has actually occurred, $<B|A>^*$ would mean how probable is A given that event B has occurred. If I hear the phone ring, has it actually rung? That is the meaning of $<B|A>^*$. If the phone has rung, have I heard it? That is the meaning of $<B|A>$.

Their answers can only come when they are multiplied together, $<B|A>^* <B|A>$. But now we use the quantum rule that says that to asterisk is to reverse. Thus $<B|A>^*$ becomes $<A|B>$. This *is* the given hearing of the phone and the probable ringing of the phone. Then the multiplication of the two "questions" produces an answer, $<A|B><B|A>$, a verification, a "handshake" across time that indeed the phone rang (event A occurred) and I heard it ring (event B).

$$<A|B> <B|A> = <B|A>^* <B|A> = \text{the probability that } B \text{ follows } A.$$

PART FOUR

Ψ*Ψ

NEW FRONTIER OF THE MIND

The only game in town these days is making matter conscious. It already is conscious but not all of matter knows this. Consciousness is trapped into states of unconsciousness. In this, the final part of the book, we examine how the universe we know *is constructed from phase harmony brought forward by the superposition of holographic layers. Consciousness is seen not only from the viewpoint of qwiff popping but also as a superviewer who, from the future standpoint, is threading and pulling us (that is, itself), as if we were all threads of space-time fabric, into the weave of an infinite-dimensional hologram made in layers — like the finest Greek filo pastry — called parallel universes.*

The division of these parallel universes is seen to split into two parallel planes, the mental and the physical. Together they constitute the new psychology: quantum psychodynamics — *the evolution of consciousness.*

9

Voyaging to Other Worlds

We should like now to comment on some views expressed by Einstein. Einstein's criticism of quantum theory ... is mainly concerned with the drastic changes of state brought about by simple acts of observation (i.e., the infinitely rapid collapse of wave functions). ... At another time he put his feeling colorfully by stating that he could not believe that a mouse could bring about drastic changes in the universe simply by looking at it.

However, from the standpoint of our theory, it is not so much the system which is affected by an observation as the observer, who becomes correlated to the system. ...

Hugh Everett, III
The Theory of the Universal Wave Function

Drawing by R. Chast: © 1982 The New Yorker Magazine. Inc.

Consciousness and Parallel Worlds

It is 80,000 B.C. and you are there. It is time for foraging. Up ahead you come to a familiar landmark, but the scene has somehow changed. Perhaps it is the light on the tall grass. Even worse perhaps there is a dreaded bear about to pounce. Last time out you went to the right. Or was it to the left? You aren't sure. You must choose — the lives of your whole family depend on making the correct choice. You go left. You sense the danger as you move hurriedly along what appears to be a familiar track. Finally you reassure yourself. You are on the correct path. Suddenly, just as you relax, the great bear pounces. A sharp pain is all you remember as everything goes black. The scene is over. You are dead.

Hold it. Stop the action. Let's run that by again. Back to tall grass. But this time you go right. As you move cautiously along the path, a path that you cannot remember, you have a tingling feeling. It is somehow familiar and yet it is different. Ahead you see what you are searching for. You gather the firewood and return to your family. The scene ends. You have survived.

It seems cut-and-dried. Either you survived or you didn't. Right? Wrong. You survived and died back there on the prehistoric veldt. You took both paths. Yet you did choose a single path. You had to. How did you do both? The answer to this seeming paradox, indeed the realization of such a question, is a product of the "new physics." As fantastic as it may sound, the "new physics" called quantum mechanics posits that there exists, side by side with this world, another world, a parallel universe, a duplicate copy that is somehow slightly different and yet the same. And not just two parallel worlds, but three, four, and even more. No less than an infinite number of them. In each of these universes, you, I, and all the others who live, have lived, will live, will have ever lived, are alive.

Where is this universe? It is literally the universe around the corner. You are entering it now; indeed, you have been doing so, perhaps unconsciously, ever since time began.

In Part Four we shall explore how this radical new idea, the parallel universe, or many worlds, interpretation of quantum physics, plays a role in our everyday lives and, in fact, actually constitutes our everyday lives. Even more startling, I hope to show that consciousness, as we presently understand it through the traditions of Freud, Jung, Buddhist meditation practice, and others, can actually be synthesized into a new world view. Through it we will gain a firmer perspective on why there is suffering — how it arises naturally out of fundamental acts of consciousness, acts that indeed "create" the universe in which we all live. These fundamental acts separate our little pieces of consciousness from the mainstream, drops from the ocean of the ever-present infinite universe of all possible parallel universes. These acts, these senses of "knowing" something, are at the root of all suffering. In each of these acts is born the greatest illusion of all, the illusion of "I." This ever-present "I-ness,"

229

this constant desire to turn "I-ness" into your "highness," is the only magic trick in town.

The key to understanding parallel universes is the recognition that time only appears to be a barrier to experience in a single universal layer. Overall, encompassing all layers of the multiverse brings to light the holographic mechanism that makes experience what it is on any one layer. Each layer contains a pattern of space-time trajectories. When these layers are superposed on each other, as if they were transparent sheets, the overall pattern appears. This overall pattern is the universal hologram. Again time is not real; it is not observable in the same sense that spatial location or momentum is observable. Since time is not an observable, what are we observing when we say we are observing time? The answer is, a memory sequence. There is no time; there are only memory sequences. Memory sequences are what we call time. These sequences are recordable in arbitrary ways. Thus what we call the past is only a matter of record. There is no such thing as *the* past; there are only memory sequences that are somewhat related to each other.

I was led to the work on parallel universe theory by Hugh Everett, Bryce DeWitt, and Neill Graham. These workers reinforced in my mind the importance of memory sequences. If one were to look at and count all the memory sequences predicted by quantum mechanics, one would find that some sequences occur more often than others. Could the simple numerical fact that sequence 1 occurs more often than sequence 2 have any bearing on the mind? If we assume that the mind, like the mind of God in Bishop Berkeley's philosophy, is and exists, then all sequences are remembered. Sequences are memory. The numerical superiority of sequence 1 over sequence 2 meant that more people would remember 1 than 2. If 1 was occurring in your own mind, then you would remember 1 rather than 2. In this sense, all factual data are simply a matter of vote, with the majority ruling.

In courtroom scenes, for example, the truth is not taken to be known. It is agreed on by jury vote. In a similar sense your own past experiences are decided by your own internal jury in consideration of several possible memory sequences. Legal facts are never presumed; they are arrived at by the lengthy processes of jurisprudence. This ultimately means a polling of the jurists. The paralleling sequences, the ones that appear the most common of human experience, are the ones we notice the most, because these sequences are relatively unchanging.

Brainwashing may occur through voting in this manner. The brainwashed individual is subjected to a powerful group or other individual consciousness. His votes are tallied along with the outside source's votes. And still the majority wins, however that majority is decided on.

We could say that there is consciousness in every possible memory sequence. If there are a number of such parallel consciousnesses, meaning a number of resonating or kindred spirits, good vibrations, etc., the mere fact of the sequences being the same would lead to a form arising between them.

This form would have the good fortune that in time — which in quantum mechanics means the internal transformation of Hilbert space onto itself — there would be no changes to first order.† The world would appear to be the same, nothing would have changed. This is the genesis of the form of habitual response to change, to leave things as they are. What is coming is the same as what has already passed.

Other memory sequences are just as real, but because there are fewer of them, transformation in Hilbert space tends to cause changes in them that eventually wipes them out. Only those sequences that are more numerous tend to be reinforced through Hilbert space transformation.

This transformation can be envisioned as a sphere undergoing rotation. But this rotation should not be considered to be time. Time arises from these rotations, but time isn't an observable. When I first encountered this, it meant to me that there had to be a quantum physics of mind: thought, feelings, intuition, and sensation — the four functions specified by Carl Jung. Jung had placed these functions in antipodal positions, with intuition opposing sensation and thinking opposing feeling. In Jung's operation a person who felt couldn't really think, for example.

"Lucid" Dreaming and My Visit to a Parallel Universe

It happens from time to time, usually when I am not expecting it, often after a sleepless period of tossing and turning. The first time it happened was in the fall of 1973. I was a visiting professor at Birkbeck College at the University of London. I had retired about 10:00 P.M. after returning from my office at Birkbeck. At 2:00 A.M. I was awake, my mind was filled with physics equations — something to do with parallel universes, other worlds nearly exactly like our own, but somehow different.

Quantum mechanics opens the door to such ideas, and I had been fortunate enough to be at Birkbeck where John Hasted and David Bohm were cochairing the physics department. Hasted was beginning to investigate parallel universe theory to explain some weird paranormal effects that he had observed in his lab. Bohm had been working since the early fifties at the roots of quantum physics, like a patient gardener. Both men must have had their influence on me that night.

I went to the dining table in my apartment so that I wouldn't disturb my sleeping housemate, Nancy. After feverishly writing down a series of seemingly indecipherable hieroglyphics on my paper pad I felt a deep sense of satisfaction

---

†By order I mean magnitude. When a mathematical function is orderly, it is called analytic and can be represented as a power series, a sum of terms. The first term in the sum is called the zeroeth order term, the second term is called the first order term, and so on. Each term in the sum is smaller in magnitude than the term preceding it. A change in sequence can be represented by an orderly or analytic function. Thus the terms higher than the first order can be neglected when compared in magnitude to their preceding terms.

and grew drowsy. I went immediately to bed and fell asleep.

Now when I say I "fell asleep," I mean more than you may think. I felt myself falling down a deep and dark well or tunnel. Yet every so often I would stop falling and find myself involved in a scene, as if I were an actor suddenly appearing on a stage. These scenes just appeared and I was enmeshed in them. I was not just an observer but was actually "there." Quickly the scene would change and I would find myself in yet another scene, entirely different from the one I had just left. These scene changes happened so rapidly that I felt I was descending from one layer of the universe to another, slipping through time and space just as a small pebble slips through the woven mesh of a fabric. As I descended I became more and more aware that I was dreaming. It was dawning on me that I was both snuggled cosily in bed and slipping through space-time in a dream of uncanny proportions. It was as if my awareness were split in two. To my great surprise I was conscious that I was asleep. What a contradiction! How can you be asleep and conscious at the same time?

Next I found myself awakening, but I was shocked to discover that I had not actually awakened at all: I was dreaming that I was awakening and I knew it! No sooner had I realized that I was still dreaming than I would awaken once more from the dream to dream that I was awakening once again. It was like ascending through a set of Chinese boxes: As soon as I was out of one box I found that I was inside of another, still larger box. I soon realized that I was in control of my dream. I could awaken for real or I could descend to any universe layer I wished and experience my dream consciously. I then decided to explore and instantly found myself in the strangest room that I had ever been in.

The room was shaped like a large cylinder and appeared to have a dirt floor. When I looked up, I saw a clear blue sky shining through what seemed to be an open roof. I found myself standing next to the room's outer wall and began to feel its texture. I was amazed to notice that I could feel the wall and that it felt neither cold nor warm to my touch but instead had a somewhat rough texture, like coarse woven fabric or basket weave. Then something quite strange happened to me.

I noticed that I was rising or floating upward in the room and immediately felt a sense of panic. This anxiety halted my rise and I descended to the floor once more. I "remembered" myself sleeping comfortably back in my earthly abode and breathed a sigh of relief. With my relaxing I immediately began to rise again toward the open roof. Again I felt fear and began to descend. With this new knowledge I experimented with rising and sinking and noticed that all I had to do to descend was feel fear. To rise all I had to do was relax. I was just getting used to my new environment and had ascended fairly close to the blue skylight when I sensed the presence of another person in the room.

Looking down I saw below me the "caretaker," a kindly but blurry-looking old fellow. My vision, I discovered, was as nearsighted as it was normally on earth. The "caretaker" announced himself to me and jovially said, "Hello. You must be new here. Come on down and I'll show you around."

Now, when I say he announced himself to me, I don't mean that he spoke. I just heard him in my mind. I couldn't describe his voice. I wasn't even sure if he was a "he." It was a kind of instant thought communication. I thought and he heard my thought. He thought and I heard his communication. I heard no sound, but I sensed his words as clearly as if he had actually spoken. And he sensed my words in the same apparent manner.

Next we walked side by side out of the room through a nondescript doorway. I found myself walking with him silently and had the feeling that I was in a quiet, beautiful countryside of rolling hills. This is how the surrounding scenery appeared to me. I say I "felt" this was the case because my feelings and my visual sensing of the surroundings were somehow the same. What I felt matched what I saw and vice versa. This is difficult to describe in words.

We continued to walk around and I sensed a great relaxation and peace. The sky was blue and cloudless. There was no sun anywhere. The grass was greener than any grass I had ever seen. I soon noticed that there was a low brick wall, perhaps three feet high, weaving through the hills and greenery. Soon I "heard" voices and saw a large group of people just ahead of us. I realized that the silent "caretaker" was leading me to the people who were sitting comfortably on the low wall and the grassy areas it enclosed.

My entrance into the group stirred no response. I was just another person there. I felt as if I had come to a picnic and yet I noticed that there was no food in sight. And still no one was paying me the slightest bit of attention.

I then began to look around at the faces of my new associates. I must point out how unusual all this was to me, because at any moment I could "remember" myself sleeping in bed at home in Shepard's Bush, London W.6. "I" was where I was and "I" was home at the same time. This experience of remembering was exactly the same as when you think back to a past experience, the only difference being that in waking consciousness you can't "return" to your memory. In my altered, or lucid, dream state I not only remembered my sleeping self, I knew I could return any time I wished to.

The reason I was thinking about going home was the bizarre physiognomy I was suddenly gifted with. I merely had to look at a face, any face, and I "saw." More than seeing, I knew. The facts of the personality were an open book to me. I merely looked at a face and it would undergo a series of transformations, each change revealing a new fact. I couldn't look too closely because, frankly, I was frightened by what I saw. On every face was great sadness and pain. The faces were normal when looked at quickly, but when examined for any length of time they became grotesque masks with great striations of contorted pain lines, hideous peelings of unfolding skin layers, and throbbing nerve threads all pulsating on raw skin.

Suddenly I realized where I was and announced to myself, i.e., thought to myself, I was on the astral plane of suicides. These people had committed suicide on earth and were waiting to reincarnate — to return to earth and be

233

reborn. But there was a slight problem. In order for them to return they had to be acceptable to all the "normal" nonsuicidal souls they will share a body with. That is why they were here: to await humanity's decision.

Each of us is a universe of souls, not just a single soul journeying from here to Timbuktu. As the Buddha taught, we are all questions of compromise. Each of us is a universe of past lives, and some of us living now owe a debt of gratitude to the others for allowing us to live again. These suicides were the astral-level component, the parallel-universe level of reality, of past failures in life. We all have in us the lives of past failures, murderers, rapists, saints, and sinners.

This realization appeared to me as a thought, but I had made a mistake. I had thought to myself, not realizing that my thoughts were open books to my fellow "travelers." And even worse, what in all hell was I doing there in the first place?

Just then I noticed "her." She was sitting on the wall facing me and, gulp, she was looking directly at me and smiling. I heard her reply, "Oh, you know where you are? Who are you? Where do you come from?" She approached me in an overly friendly manner. I boasted, "Yes, I know where I am and I can return home any time I want to."

"You can, can you?" she asked with great interest as she came close to me. I was getting frightened. This was my first trip and I didn't know what danger I might be in by my just being there. Then I looked at her eyes. I don't quite know how to describe what I saw, but her eyes began to spin. They appeared to me as rotating pinwheels of spiraling colors. She was now too close for comfort. I knew then that I had to leave and I exercised the "leaving ritual," the only one I knew would get me out of there fast. I yelled bloody murder.

I awoke in bed next to Nancy, and this time for real. It must have been past four in the morning and Nancy wasn't too happy to have me just pop up in bed talking a blue streak. I not only was wide awake, I was fully conscious and quite lucid and gregarious. Loudly I said to her, "Nancy, wake up. I must describe this dream to you now before I forget it." Nancy, hardly believing her eyes or ears, was rudely being shaken from a deep sleep of her own. And dazed but understanding, she listened to the story of my voyage.

It is very important to realize that this "dream" was not just an ordinary dream. I was fully conscious not only during it but in the transition from the astral plane to my bed. My yelling was soundless in the astral realm but gradually became real sound in the physical plane of the bedroom. There was no need for coffee. There was no sleepiness, nor did the dream fade from memory as I became more awake (as most ordinary dreams do). It was simply a matter of recalling actual events in the same manner as you would recall events of the morning over an afternoon lunch.

I hadn't been asleep and I wasn't simply lying in bed and daydreaming.

A few weeks later Nancy, her friend Ann, and I went to a Friday evening service of the Druid order presided over by the then chief Druid, Dr. Thomas Maughan. Dr. Maughan, now deceased, was a remarkable man. His service was open to any who wished to attend and he always had an open ear and mind to "voyages" such as mine. After the service Nancy asked me to speak up and tell Dr. Maughan about my dream, and I did. Maughan listened attentively. When I finished, he looked at the group of attendees and reaffirmed that this was no ordinary dream. He asked me what I did for a living. I told him that I was a visiting professor of physics at Birkbeck College. Then, astonishingly, he admonished me for not being more attentive to details when I was there. "You weren't a too careful physicist. You should have talked to the woman, asked her her name, gotten her address and phone number," he scolded.

And he was right. I was so struck that I had actually gone to the astral level and had had the experience that I failed to take it into account as an experience worthy of physical laws and subject to the same scrutiny as any other physical experience. My own scepticism had defeated my acceptance of its reality. To my rational mind "it was only a dream."

Over the passing years I have had several similar dreams. However, I have never returned to the astral plane of suicidal souls.

Lucid Dreams and Real Holographic Images

In an earlier chapter I discussed the holographic model of the brain. I had mentioned that when a new light wave was shined through the recording medium, two images were generated, a virtual image and a real one. I then explained how the virtual image appears to be coming from the original space behind the hologram because the new light wave passing through the hologram is "screened" by the holographic pattern in such a manner that it looks like the light waves coming from the original object.

On the other hand, these light waves also contain additional information. If left alone, that is, if nothing is put in their way after they pass through the hologram, they will come to a focus in space, forming a real image of the object in the focal space. When seen, this image will appear to be floating in space. The reason that this is a real image and not a virtual one is that the light waves actually come from the real image when they are seen. On the other hand, light waves appear to come from a virtual image but actually do not.

In usual applications of holography little attention is paid to the real image because the image is three-dimensional and needs special care in order to be seen. Placing a screen at the focal space of the image only picks up that part of the image that is intercepted by the flat surface, a slice of the image. The virtual image appears three-dimensional but is actually an illusion of three-dimensionality produced by the two-dimensional hologram. The virtual light waves appear to be diverging from a real object in back of the hologram, but are actually being formed by the flat hologram. The real light waves, on the other hand, are being focused by the hologram, and the focal points are the points of the image in space in front of the hologram.

Of what use is the real image in the brain hologram? I suggest that these real images can be seen when we are asleep. These images constitute the lucid type of dream I discussed in the previous section. To see how these dreams arise we need to return to the notation used in Chapter 6. The brain is again pictured as a recording medium. The information waves corresponding to the waves of light coming from the object are denoted as $[IW1]$, $[IW2]$, etc. Each of these waves contains information about the object in question. These could be bits of computer data, optical pixel information, bits corresponding to sensations, etc.

These information waves need to contain both amplitude and phase information. For quantum wave functions — qwiffs — the amplitude says something about how probable it was that the event corresponding to the information had occurred. The phase corresponds to the action involved in transmitting the information to the recording medium. In optics the phase corresponds to the number of wavelengths a light wave travels from the illuminated object to the recording medium. In the brain the phase will depend on what kind of wave is making the record. If it is a qwiff, the phase will depend on the path of action (see Chapter 3, page 103). Thus these waves are complex numbers. Their complexity makes them useful and offers the surprising explanation of how lucid dreams differ from sensings of outside experiences.

Suppose that the cortex contains recorded information. This record was made from the interference pattern produced by the reference wave $[RW]$ interfering — adding together — with the information waves from past events. The mathematical form for all this involving information wave 1 is:

$$[RW] + [IW1]$$

Multiplying by the complex-conjugate sum

$$[RW]^* + [IW1]^*$$

produces four terms:

$$\underset{\text{term 1}}{[RW]^*[RW]} + \underset{\text{term 2}}{[RW]^*[IW1]} + \underset{\text{term 3}}{[IW1]^*[RW]} + \underset{\text{term 4}}{[IW1]^*[IW1]}$$

Terms 1 and 4 are composed of complex numbers multiplying their complex conjugates, and thereby they have lost all phase information (see Chapter 6, page 158). Terms 2 and 3 are potentially able to reproduce the original information wave because this phase information has not been lost in the multiplication with the reference wave.

These four terms compose a record in the cortex. To see the record, i.e., replay the recording, another wave $[NW]$ is "shined" through the cortex. The cortex acts as a transmission filter for this new wave and thus in effect multiplies the recording by the amplitude of the new wave. This gives:

$$[NW][RW]^* [RW] + [NW][RW]^* [IW1] + [NW][IW1]^* [RW]$$

term 1   term 2   term 3

$$+ [NW][IW1]^* [IW1]$$

term 4

Now, what is observed depends on the phase and amplitude of the new wave. If the new wave is the original reference wave or similar to it:

$$[NW] = [RW]$$

then terms 1 and 4 will appear as a "clouded" reference wave. But term 2 will have

$$[RW][RW]^* [IW1]$$

its first two coefficients, $[RW]$ and $[RW]^*$, multiplying each other and canceling their respective phases, leaving the information wave, $[IW1]$, alone. Term 3 will not survive the onslaught of $[RW]$ because when the new wave, $[RW]$, multiplies the third coefficient, $[RW]$, no reference wave information is wiped out. Instead these two coefficients together mask the information contained in $[IW]^*$.

The result of this is a replay of the virtual image, an appearance of the image in the outside world. In order to generate this image the reference wave must come from the outside world. This could be a simple light wave reaching the eye or a sound wave hitting the eardrum. Any outside source of sensation will stimulate the record, producing the image. The superposing of this image together with the actual imposition of outside information constitutes the holographic comparison of a memory with the new outside source of information.

In this way the continuing newness of experience appears redundant to our minds. We don't ever see what is "out there" as distinct from our own memory record. Such is the physics of boredom and habituation. Everything is new but we always see our records. Our brains work in this manner to ensure our survival. We want to make sure that threats are recognized. To do this we need our records. Our brains are like music lovers who insist on carrying tape recorders with them to the concert. When a piece is performed, the recorder is turned on and plays the same tune as the orchestra is playing. Or nowadays, baseball fans carry portable TVs with them to the ballpark and watch the game they are watching "live."

However, when we sleep and manage to reach the altered state of consciousness known as a lucid dream, a different tune is played back.

Again a new wave is generated, only this time it is not an external wave. The new wave is instead the complex-conjugate reference wave

$$[NW] = [RW]^*$$

Terms 1 and 4 will appear as "clouded" as before, but term 3 will have

$$[RW]^*[RW][IW1]^*$$

Its first two coefficients, $[RW]^*$ and $[RW]$, multiply each other and cancel their respective phases, leaving the information wave $[IW1]^*$ alone. This time term 2 will not survive the onslaught of $[RW]^*$ because when the new wave, $[RW]^*$, multiplies the third coefficient, $[RW]^*$, no reference wave information is wiped out. Instead these two coefficients together mask the information contained in $[IW]$.

The result of this is a replay of the real image, an appearance of the outside world.

This appearance is not the real outside world but the world of our dreams. Since the image is a real one it is formed by waves coming to a focus, in contrast to the virtual images, which appear to be coming *from* a focus, diverging from an apparent source. If there is a "viewer" where these waves focus, that viewer will be bathed in the scene and the scene coming to a focus will "contain" him. In this way the dream experience will appear "lucid."

It is important to realize that this dream state is caused by the internal generation of the complex-conjugate reference wave $[RW]^*$ and not the external reference wave used to make the original recording, $[RW]$. Any complex-conjugate wave is a kind of mirror to its wave; for example, $[IW1]^*$ is a mirror wave to $[IW1]$. If, say, $[IW5]$ was a wave traveling from left to right, $[IW5]^*$ would be a wave traveling from right to left. If the wave was a spherically expanding wave, the *-wave (star wave) would be a spherically collapsing or imploding wave. What exploded in the source and appears as exploding in the virtual image will appear to be imploding in the virtual image.

If we now look at a series of records recorded in real-time sequence 1, 2, 3, 4, etc., producing the total memory record

$$[RW]^*[RW] + [RW]^*[IW1] + [IW1]^*[RW] + [IW1]^*[IW1]$$

| term 1 | term 2 | term 3 | term 4 |

$$[RW]^*[RW] + [RW]^*[IW2] + [IW2]^*[RW] + [IW2]^*[IW1]$$

| term 5 | term 6 | term 7 | term 8 |

$$[RW]^*[RW] + [RW]^*[IW3] + [IW3]^*[RW] + [IW3]^*[IW3]$$

| term 9 | term 10 | term 11 | term 12 |

$$[RW]^*[RW] + [RW]^*[IW4] + [IW4]^*[RW] + [IW4]^*[IW4] \ldots$$

| term 13 | term 14 | term 15 | term 16 $\ldots$ |

then when the reference wave [*RW*] is transmitted through once again, it picks up terms 2, 6, 10, and 14. This occurs because multiplying [*RW*] by each of the terms above fails to destroy the reference wave "drone," which is its constant and repeating phase information in all but these terms. The drone reference wave "drones out" like a loud bagpiper all of the terms that do not contain a single [*RW*]* term to cancel out the drone phase. All the selected terms do contain the needed [*RW*]*, and when [*RW*] multiplies them it leaves the remainder of each term free as a carrier wave of information. (Actually, ordinary radios cancel the station wave frequency in a similar manner. The high-frequency station wave is canceled by the action of a rectifier, which in effect leaves the message wave free to reach the loudspeakers.)

The result is that what is witnessed when the reference wave is shined through the medium is:

$$[RW][RW]^* \; ([IW1] + [IW2] + [IW3] + [IW4] + \text{etc.}$$

term 2   term 6   term 10   term 14

with the other terms forming a background of "noise." Since they are added together, these waves form a wave superposition with a resulting interference pattern.

In this manner, when we see an object "out there," we not only see it but we replay all the previous information connected to it through past information recordings.

Not only do we carry tape recorders to concerts but we also insist on playing back every tune that even remotely sounds like the concert performance!

When we sleep, something else takes place.

If we happen to replay the right *-wave instead of the reference wave — we shine [*RW*]* instead of [*RW*] through our brain caps — the above terms are themselves droned out and the real images of the above experiences replayed instead:

$$[RW]^* \; [RW]([IW1]^* + [IW2]^* + [IW3]^* + [IW4]^* + \text{etc.}$$

term 3   term 7   term 11   term 15

Here the lucid dream starts, where you experience the world at the focus of these star waves. In your dream you will be awake and will experience sensations and feelings brought on by the actual experiences "learned" in the outside world and that produced the information waves in the first place.

Since these waves are star waves, they are the mirror images of your experiences in both time and space. Where term 3 occurred before term 7, in your dreams they will appear in a time-reversed order. This is akin to a direct experience of your id, where, as Freud pointed out, "logical laws of thought do not apply . . . there is nothing corresponding to the idea of time. . . ." You do experience time in your lucid dream, but it is not real time; it is what your previous information waves told you time was.

By selectively changing the reference wave, as we do when we simply look out at the objects surrounding us, we experience different sensations and feelings. We have different thoughts and intuitions. In Chapter 10 I will discuss how the parallel worlds theory of quantum physics offers an explanation of how we come to have these internal experiences as a result of our external experiences. We will see how the qwiff pop of consciousness can be avoided, provided we allow an infinite number of parallel universes to coexist. In Chapter 11 I shall carry this further. Not only does the parallel worlds theory explain and rid us of the qwiff pop (not the qwiffs; the waves and star waves remain), it also provides a means to explain our internal mind world and enables me to show you how your waves and star waves are recorded as memories causing feelings, thoughts, intuitions, and sensations — all you need to feel alive.

10

The Game of Consciousness and Parallel Universes

Verily, there is a realm where there is neither
the solid nor the fluid, neither heat nor motion,
neither this world nor any other world, neither sun
nor moon . . . There is, O monks, an Unborn,
Unoriginated, Uncreated, Unformed. If there were
not this Unborn, this Unoriginated, this Uncreated,
this Unformed, escape from the world of the born, the
originated, the created, the formed, would not be
possible.

Sākyamuni Buddha [quoted in "Vanishing Magician-Spectator,
Rabbit, and Hat," by Claudio Naranjo, M.D.
Reflections of Mind]

What Are Parallel Universes and Why Are They Necessary?

Up to now I have described how consciousness enters the physical plane of existence. This entrance is a rather dramatic disturbance. It is a disruption of the smooth, anticipated, organized movement of any object from any place at any time to any other place at any other time. It is a sudden and explosive qwiff pop in which what has gone before is completely lost to memory. There is an abrupt event, which is immediate experience. In this instantaneous "collapse" of the qwiff "bubble" the reality, which is totally described as a qwiff flow, becomes a single particulate event occurring at one instance of time and at one specified location in space. This action of consciousness cannot be explained by any scientific theory known today.

It is an acausal action and is therefore quite mysterious. There is no reason for it to occur. This unreasonableness of the action of consciousness is disturbing not only in its activity but also in the psyches of quantum physicists, who are those elements of consciousness created to explain this action.

Quantum physics is a beautiful theory. It explains as no other physical theory can such mysterious behavior as:

- How a wave can be a particle
- How a particle can "tunnel" through obstacles
- All of chemistry
- The structure of things
- How light interacts and alters matter
- Why the stars shine
- Light
- Matter
- Energy

and lots more. So far there is no theory that surpasses it.

Its major fault is its interpretation. There is an accepted interpretation called the Copenhagen interpretation. It says that observation requires a macroscopic, or large with respect to atomic behavior, instrument to register the events we call our experiences; that, as a result, all interpretations of what is going on are "doomed" to classical conceptualization. Consequently these observations are disturbing because we, by the very nature of our classical prejudices — brought on by the necessity of using macroscopic instruments — do not possess an adequate language of that description. Furthermore, we will never be able to gain or possess that language because the very act of thinking about it means that we have already passed into our own classical measuring instrument — our human brains and nervous systems — to accomplish this act.

An atom may not be at all what physicists say it is, no matter how accurate that description, because an atom is a classical concept. Thus all the weirdness of quantum physics is present only because we are doomed to ask the wrong questions all the time. Our thinking is inaccurate because we use inappropriate tools of thought, words such as *particle, wave, energy* — which, although perfectly acceptable at the level of baseballs, oceans, and calorie counts, are not even close to what is going on at the atomic level.

However, this is the best that is possible. Uncertainty must exist if we think of atoms as things (and there is precious little else to think of if atoms are not things). Once we allow the possibility that atoms, electrons, and all that can be built up from atoms and electrons are *not things,* the doorway to imagination opens and all hell can break loose.

This book is the result of my thinking about the discoveries of quantum physics and its relationship to observation, the mind, and the recognition that things are not things. A thing is only a thing at a level of description where the results of observation appear to play an insignificant role. A thing is a thing if it occupies space. Even a green dragon with white polka dots that flies and eats purple people is a thing because it occupies a "space in my mind."

I have brought out the concepts of qwiffs and qwiff pops to describe the physical and mental worlds as playful alternatives to the rigid mechanical descriptions used in traditional physics. As such these newer concepts introduce a quirky kind of humor into the game of the universe. This game also requires something outside of the physical universe acting capriciously in it — namely, a consciousness that pops the qwiffs.

This description is flawed in one way: It is difficult to see how consciousness can also be a qwiff. To pop a qwiff and be one is self-defeating, to say the least. Somehow, as I have inferred throughout the book, consciousness acts self-consciously in its qwiff-popping activity. To accomplish this task, I ask that the qwiff "interact" with itself, multiply itself in some manner, producing as the result the relative probabilities of the events predicted to occur by quantum theory.

This is no surprise. The calculation of probabilities requires this multiplication (as described in Chapters 6 and 8) as part of the accepted theory. All physicists accept that probability is calculated in this manner.

I have attempted to go further. I stated that this multiplication not only occurs in the minds of the calculating physicists but also as a process in the physical world. I further stated that this multiplication is the result of the qwiff "doubling back" on itself as a reflection from a possible location in space or as an abstract movement within a mind-space called the Hilbert space.

There is no time barrier in this. The flow can go from now to the past or from now to the future.

The mind *now* holding the abstract event "in mind" reaches out into space to the event *then* and then doubles back from space to the mind once again. This out-and-back flow "creates" both the physical event and the knowledge imprint of that event as probabilities. Repeating the process over and over again (outside of time or simultaneously) alters the odds until, finally, we accept as real what we think we have experienced. In this abstruse manner, realization is sensed to occur.

There is an important connection between the quantum wave function, or qwiff, and the quantum-mechanical probability amplitude, or quamp, that must be understood. A quamp is a question composed of a bra, $<$|, and a ket |$>$. For example, the quamp $<b|a>$ is the question: If a is true, then is b true? Remember that a ket is an abstract *initial* state of a system and a bra is an abstract *final* state of the same system. If either a or b is the state described by the observable "location in space," then the quamp is a qwiff. If the a in ket-a is a spatial location, it is a qwiff flowing forward through space and time; and if the b in b-bra is a location in space, it is a qwiff flowing backward through space and time.

Thus if one is dealing with kets, it is always possible to switch "representations" to qwiffs, and vice versa. In the next chapter I will discuss this connection in greater detail. For our purposes here, it is only necessary to note that the transformation from qwiff language to quamp language is essentially no different from the translation of one "tongue" into another. So in all that follows in this chapter I shall use qwiffs, quamps, bras, and kets interchangeably.

Also, I throw out our insistence that time is only now, that existence somehow disappears into past, and that future never *is* but only will be. All events are in a supernal sense "now."

Human behavior is a complex set of space-time operations whose outcomes are experiences. Our operations, called observables in quantum physics, affect the outcomes. Observables are sandwiched between the actions of the outflowing future-seeking qwiff and the back-flowing past-seeking qwiff. The result is the production of an average, or expectation, value for the observable. This result *is* what is expected as the event! Indeed the actual measurement of the value, when observations are carried out in a large number of similarly prepared systems, is usually close (within a standard deviation) to the predicted or expected value. This is much the same as a crapshoot with a million dice, for which the average, or expected, value will be 3.5 (i.e., $[1 + 2 + 3 + 4 + 5 + 6]/6 = 3.5$). I showed this mathematically in Chapter 8.

This average is not, however, the qwiff pop. For that to occur the qwiff must suddenly reduce itself and become one of its possibilities. That is the average

$$<I|A|I> = \text{SUM (over all } i, \text{ from } i = 0 \text{ to } i = \text{infinity)}$$
$$a_i <I|A_i> <A_i|I> \text{ —pops} \rightarrow a45, \text{ e.g.}$$

In other words, from all of the infinite values to be realized, only one value appears, as for example $a45$. The sum jumps to one value. Now this jump or sudden collapse changes the original ket-I to ket-$A45$:

$$|I> \longrightarrow |A45>$$

in the process. Now, ket-$A45$ is not special. It could have jumped to ket-$A1003$ or ket-$A2$. The ket "chosen" is chosen probabilistically. We know that such a jump occurs because when the expectation is calculated again using $|A45>$ instead of $|I>$, and when the measurement is repeated using the same A observable on the same system, the value $a45$ recurs.

Here we use the corollary to Law III (Chapter 8, page 200), called orthogonality. Since $|A45>$ is unique:

$$<A45|A_i> = 0, \text{ for all } i \text{ not equal to } 45$$

i.e., ket-$A45$ is orthogonal to all other kets. This means

$$<I|A|I> = <A45|A|A45>$$
$$= \text{SUM (over all } i, \text{ from } i = 0 \text{ to infinity)}$$
$$a_i <A45|A_i> <A_i|A45>$$
$$= a45.$$

The ket-$A45$ now is the new ket-i for this recurrence:

$$<A45|A|A45> = <i|A|i> = a45$$

Without this recurrence it would be impossible to have any "memory" or repeatable observations at all. This jump is sometimes referred to as a *state preparation* and the next measurement, in which no jump takes place, is called the *actual measurement.*

If one is "forgiving" in the process of measurement and doesn't attempt to discern the difference between $a45$ and the expectation value $<I|A|I>$ in the first place, then no collapse is observed to occur. Ket-I remains ket-I and does not jump to ket-$A45$. The state preparation really doesn't prepare anything new.

It is only in the process of fine discernment or careful measurement that a jump to a new state can take place. Indeed, herein lies the hope of transformation. It is the act of observation that "caused" the jump.

This act is the most disturbing to science. It has eluded all attempts to fit it into a mathematical scheme. When I say this I don't mean that one cannot compute anything at all. I mean that the prime goal of a mathematical scheme is to provide an algorithm that enables a prediction of an observation to be made. Usually this implies a deterministic physical theory, one that says what will occur when and where. Since quantum mechanics is only deterministic in its predictions governing kets, bras, qwiffs, and quamps — i.e., mathematical forms in Hilbert space, not physical events in our four-dimensional space-time continuum — and since these very same kets, bras, qwiffs, and quamps must change radically and indeterministically, that is, they jump whenever an original observation takes place, many physicists believe that quantum physics is simply an incomplete theory.

What it would take to "complete" quantum physics has been investigated rigorously ever since its formulation in 1925. Various schemes have been proposed involving the inclusion of all sorts of "nonlinear terms" in the mathematical algorithm that does predict the evolution of kets, bras, qwiffs, and quamps. None of these schemes proved satisfactory, however. The frustration at being unable to predict the collapse of a qwiff or the jump of a ket led Hugh Everett III, in his 1957 Ph.D. thesis with Princeton University's physics department (his thesis advisor was the profound physicist John Archibald Wheeler) to make the radical proposition that no such jumps or collapses ever take place.

This appears to fly in the face of what has just been said. How could Everett make such an outrageous statement? Everett's thesis was a new interpretation of quantum mechanics which denied the existence of the classical, or macroscopic, realm that Niels Bohr and his Copenhagen school insisted must be there for any observation to be made. Instead Everett proposed that there was one and only one ket or qwiff for the whole universe and that this ket never jumped at all. It always followed the predicting algorithm of quantum physics, which specified deterministically how the ket continued to evolve.

Thus Everett was offering to physics a new determinism, one that was quite rigorous. Reality was saved from indeterminism when one looked at *all of reality* and not just the single immediate reality experience of one observer.

This new *reality,* however, encompassed far more than just our four-dimensional continuum. It was a reality of many worlds. Worlds beyond count. Worlds on top of worlds. Parallel worlds. Worlds that, although parallel, intersect with each other in myriads of designs, whenever interactions occur. The *one ket,* the big ket-I, $|I>$, decomposes into orthogonal kets (see page 245), $|i>$'s (i.e., kets that have no overlap). The realization of one, say, $|5>$, is the exclusion of all others, $|1>, |2>, |3>, |4>, |6>, \ldots$, just as going along one branch on a tree naturally excludes the other branches. Ket-I's branches reflect a continual splitting of the universe into a multitude of mutually unobservable but equally likely *real* worlds. In each of these worlds a human being or an instrument observed by a human being has yielded a good measurement and an unambiguous result. Thus in each world the observer in that world says that the ket jumped from ket-*I* to ket-*i* (i.e., $|I> \rightarrow |45>$, e.g.).

No such jump occurred. The ket and the observer simply interacted and became smoothly and deterministically correlated as a branch of one single ket. Ket-*I,* existing in all of its superpositions (see Chapter 8, Law IIa, " 'I' Am All My Projections"), altered the observer and "sucked" him in so that corresponding to each observational possibility there is an observer "mind state" seeing that particular observable value. The observer has been split by the relentless qwiff.

Much like a tempted Adam, who on seeing the Tree of Knowledge must taste its fruit and hence enter the inextricable state of "worldly involvement," the observer has "fallen from grace" and become part of the physical process in all of its many possibilities.

The observer in world 45 does not know of the existence of any of the other observers. As far as he is concerned the ket has just jumped. He has "awakened" to a new world because he has seen the physical "light" of the $I \rightarrow A45$ transition. To "him" a jump occurs because he cannot "feel or see" the splitting qwiff that has become his own nature. He has become many and yet feels lonely and only one.

These infinite observers are all just one observer who himself has been split as a result of his interaction with the observable system. Each of us is a multiverse unto ourselves.

For example, instead of:

$$<I|A|I> = \text{SUM (over all } i, \text{ from } i = 0 \text{ to } i = \text{infinity)}$$
$$a_i <I|A_i> <A_i|I> \text{ —pops} \rightarrow a45,$$

we have, according to Everett:

$$<I|A|I> = \text{SUM (over all } i, \text{ from } i = 0 \text{ to } i = \text{infinity)}$$
$$a_i <I|A_i> <A_i|I>$$

with no pops occurring. In each world an a_i has been seen. Each a_i is different. Each ket-A_i is orthogonal to the other kets. No ket is favored as the one that is actually observed. The ket-*I* has just split into an infinite number of parallel universes, each universe, *i,* containing a particular ket-A_i and yielding a definite and observed result, a_i.

Just this is necessary to have a determinism in the "new physics." Each and every thing is indeed not a thing, not a single solitary thing. Instead each and every thing is a universe unto itself. Even more, these things are capable of undergoing further splits, infinities multiplying infinities and then again.

What originally began as the first law of physics and consciousness, "The Mind Is One":

$$<I|I> = 1$$

is never violated. This gigantic undifferentiated consciousness is also a sum according to the second law, "The Mind Is Many":

$$<I|I> = \text{SUM (over all } i, \text{ from } i = 0 \text{ to } i = \text{infinity)}$$
$$<I|i><i|I>$$

and this multitude or, as it is often referred to by physicists, this superposition is never felt by anyone. Each of us fails to feel it because no interaction has occurred.

In the following paragraph read the word *I* as belonging to yourself. Don't think "the author" when you read "I."

Although I am a "legion," I am unaware of my multiple personalities because nothing in the outside world has caused me to become aware of my superpositional multitude. The worlds involved in the multitude remain in phase with each other. Since I have observed nothing I am the multitude of all possible observations, a unity composed of many.

But what happens when there is an observation? In the next section we will see. Again, determinism appears. We already know what you saw. Welcome to the superdeterministic, consciousness-causalistic, wave-functionalistic machine.

How Does Consciousness Fit In?

In this superdeterminism, consciousness becomes defined in a manner very different from the qwiff-popping (collapsing) or ket-jumping mechanism. However, as different as it will appear, the results of the actual awareness of psychic states, observational mind states, and memories, as well as observation of physical states — things — will be the same. It is, after all, only a matter of interpretation whether you prefer to think of yourself as an infinite, deterministically functioning quantum machine (a gigantic qwiff interacting with itself) or as a ghostlike consciousness lying outside the material universe and sticking your nose in to take a sniff and pop a qwiff.

In Chapters 1 and 2, I discussed the role of interaction. Its job is to correlate the two or more interacting objects participating in the interaction. I shall call the interacting objects subsystems.

For example, the form of interaction is, at the fundamental level of quantum electrodynamics, the photon. This particle of light is exchanged in a brief encounter. For example, one electron exchanges a photon with another (see Figure 28). The result of the exchange is that the motion of the two electrons involved is altered. The photon takes energy and momentum from one and gives it to the other. This exchange acts as the quantum-mechanical correlation between the two electrons. There is a continuous flow from the independently existing electrons to the coupled, dependently joined electron pair.

In general, any object capable of interacting is a subsystem, and its behavior falls into the bra and kets or qwiffs of quantum mechanics (see Chapter 8 for definitions of bras and kets).

In what follows we shall look at two simple subsystems to see how consciousness could result from their interaction according to the concept of parallel worlds.

> Suppose that each subsystem is represented by a simple ket. The symbol $|a>$ stands for the first subsystem ket, and $|b>$ for the second subsystem ket. And suppose that each subsystem is capable of being in one of two "states," 1 or 2. So $|a1>$ means subsystem a in state 1, while $|a2>$ means subsystem a in state 2; $|b1>$ means subsystem b in state 1, while $|b2>$ means subsystem b in state 2. Now suppose further that subsystems a and b are separate and not interacting. According to quantum physics their kets are independent of each other. Just as one computes the probability for the simultaneous occurrence of a tossed coin falling heads and a bouncing die landing with number 2 as $1/2 \times 1/6$, so one computes the ket for both subsystems existing simultaneously as $|a> \times |b>$.
>
> The kets act independently until there is an interaction between them. This interaction "couples" them so that if a is in state 1, then b is also in state 1; or if a is in state 2, then b is also in state 2. The interaction causes the initial multiplied ket state to split accordingly:
>
> $$|a>|b> \longrightarrow |a1>|b1> + |a2>|b2>$$
>
> Now the subsystems are "entangled." Where before each subsystem was independent of the other and at the same time unknowing of its own state of being, the split has altered each subsystem's unknowing state. In branch or world 1, subsystem a "is aware of" or has observed subsystem b's state to be 1. In branch or world 2, subsystem a "is aware of" or has observed subsystem b's state to be 2, and vice versa (i.e., b is equally aware of a.).
>
> Now suppose that there is a second interaction between the two subsystems. There are two possibilities. Either the subsystems continue to remain unaffected by further interaction, leading to:
>
> $$|a1>|b1> + |a2>|b2> \longrightarrow |a11>|b11> + |a22>|b22>$$

a "memory" state where subsystem *a* matches its "memory" sequence of 11 with *b*'s sequence of 11 and subsystem *a* also matches its "memory" sequence of 22 with *b*'s sequence of 22. If this is the situation, further interactions produce endless repeats:

$$|a11\ldots1>|b11\ldots1> + |a22\ldots2>|b22\ldots2>$$

but only two terms or two worlds. This continual interaction is a rote memory in each subsystem's "history." This simple repetitive pattern is a primitive form of consciousness. The continual repetition of sequences containing identical states in each world will after a while "drone" off into stoic "belief" that each world is the one and only real world. Here each subsystem has "learned" about itself from the other subsystem. This primitive consciousness is "real" in world 1 and "sensed" as possible in world 2; or "real" in world 2 and "sensed" as possible in world 1. At this level there is no distinction. The worlds are equal.

If the "other world" wasn't present, the one world would be the one entire reality. However, there would be no consciousness. There would be no hope, no expectation, no value. That dim presence of the "other world" is the dream for the world that considers itself real. And it is this dream that makes consciousness consciousness. Consciousness is the "hope" of the material world. In their primitive parallel branches the two subsystems in world 1 experience 1 as real and "dream" 2. The two subsystems in world 2 experience 2 as real and "dream" 1.

But there is more to consciousness than this.

Suppose again that the subsystems interact. However, each interaction is not the same as before. Each subsystem is not defined so that its previous "history" has any effect. The interaction allows a subsystem to change from 1 to 2. Thus a subsystem in state 1 could after an interaction make a transition to state 2, and vice versa. But if subsystem *a* undergoes a transition from 1 to 2, subsystem *b* will follow suit. So that after one interaction:

$$|a>|b> \longrightarrow |a1>|b1> + |a2>|b2>$$

However, after the second interaction each term or each world could further split, leading to:

$$|a1>|b1> \longrightarrow |a11>|b11> + |a12>|b12>$$

$$|a2>|b2> \longrightarrow |a21>|b21> + |a22>|b22>$$

Or the initial ket would become four worlds after two interactions. This would yield:

250

$|a>|b> \longrightarrow |a1>|b1> + |a2>|b2> \longrightarrow$

$|a11>|b11> + |a12>|b12> + |a21>|b21> + |a22>|b22>$

In this case two new terms are present, corresponding to the possibility that both subsystem *a* and *b* followed nonrepeating sequences even though there was a matchup where each subsystem mirrored the other.

After a third interaction there is another split, leading to

$|a11>|b11> \longrightarrow |a111>|b111> + |a112>|b112>,$

$|a12>|b12> \longrightarrow |a121>|b121> + |a122>|b122>,$

$|a21>|b21> \longrightarrow |a211>|b211> + |a212>|b212>,$

$|a22>|b22> \longrightarrow |a221>|b221> + |a222>|b222>.$

or

$|a>|b> \longrightarrow$
     1st split

$|a1>|b1> + |a2>|b2> \longrightarrow$
        2nd split

$|a11>|b11> + |a12>|b12> + |a21>|b21>$
        $+ |a22>|b22> \longrightarrow$
           3rd split

$|a111>|b111> + |a112>|b112> + |a121>|b121>$
      $+ |a122>|b122> + |a211>|b211> + |a212>|b212>$
      $+ |a221>|b221> + |a222>|b222> \longrightarrow$
           4th split

Here we see a geometric progression in which the nonthreatening interacting subsystems produce two, then four, then eight terms in the resulting ket. With each interaction the progression increases. *After thirty encounters between the subsystems there are over a billion terms in the ket!*

As maddening as all this appears, order and higher consciousness are emerging from this. To see it let's carry out the fourth split (continuing after the third split indicated above), yielding sixteen terms:

$\longrightarrow$
4th split

$|a1111>|b1111> + |a1121>|b1121> + |a1211>|b1211>$
     $+ |a1221>|b1221> + |a1112>|b1112> + |a1122>|b1122>$
     $+ |a1212>|b1212> + |a1222>|b1222> + |a2111>|b2111> +$

$$|a2121>|b2121> + |a2211>|b2211> + |a2221>|b2221>$$
$$+ |a2112>|b2112> + |a2122>|b2122> + |a2212>|b2212>$$
$$+ |a2222>|b2222>$$

If you look at these terms or worlds, you will find the following distribution:

Number of worlds with sequence of four 1's 1
Number of worlds with sequence of three 1's and one 2 4
Number of worlds with sequence of two 1's and two 2's 6
Number of worlds with sequence of one 1 and three 2's 4
Number of worlds with sequence of four 2's 1

Since the number of worlds with an equal number of 1's and 2's is greater than the number of worlds with more of one state than the other, one could assert that, if there was a crapshooting God who existed somehow outside of the two subsystems, He/She would probably choose that sequence of four interactions that led to two 1's and two 2's.

After thirty such interactions the number of terms with roughly equal numbers of 1's and 2's far surpasses any "maverick worlds" of highly unbalanced sequences.

(With thirty interactions there are 30!/15! x 15! (over 155 million) sequences of equal numbers of 1's and 2's, and over 145 million sequences containing either sixteen 1's and fourteen 2's or vice versa. As the sequences contain fewer numbers of either 1's or 2's the number of possibilities is also reduced. There is only one sequence containing all 1's or all 2's and only thirty sequences containing a single 1 or 2.)

If we disregard the order in the ket sequences insofar as when a particular 1 or 2 occurred in the sequence, we could simply collapse the above sixteen terms down into five terms where the occurrence in time of any particular state is disregarded (i.e., 1122 = 1221 = 2121, etc.):

$$|a1111>|b1111> + 4 \times |a1112>|b1112>$$
$$+ 6 \times |a1122>|b1122> + 4 \times |a1222>|b1222>$$
$$+ |a2222>|b2222>$$

Now comes consciousness and the unconscious. The third term is a superposition or composite of six worlds or six branches of the sixteen-branched treelike ket. This sheer numerical supremacy is "felt" in all of its branches as a *resonance*. This resonance is "awareness" of two 1's and two 2's. On the other hand, there is a smaller resonance between the worlds containing different populations in their sequences. They are seen relatively, from the vantage point of the overlapping majority of equally populated sequences worlds, as dream possibilities such as fear and hope. Fear and hope are the stuff that our dreams are made of. They consist of sequences in the other parallel worlds that are less likely to occur. The possibility of 1111 reflects the smaller probability of that sequence occurring. From the vantage point of the majority of the universes, where equal numbers of 1's and 2's occur, 1111 and 2222 could be the achievement of success and failure, the feeling of hope and despair or fear. These "unconscious feelings" amongst the majority are in this simple example primitive, timeless emotional states.

For the "all 1's" case there is comparatively little consciousness at all because there is no other world with which to resonate. It is likely to be a primitive, relatively unconscious mechanical world. For the "three 1's" sequence there is a weaker resonance and consequently less consciousness than the equally populated majority but more consciousness than the "double snake eyes" world.

For this to work, the "memory" of any difference between, say, a 1122 and 1221 sequence must not exist. There cannot be any causal temporality that would "fix" one sequence over another. In other words, the 1122 sequence must describe as equally temporal and with no prejudice all six sequences: 1122, 1212, 1221, 2211, 2121, and 2112. (This is essentially the same as noticing after a long drive how many right and left turns you made. You probably wouldn't remember when you made any particular right or left turn, but you would remember whether or not you made a preponderantly greater number of one kind over the other.)

Then the dream and real worlds are indistinguishable. The sense of awareness is strong. The 1122 world's dreams of 1212, 1221, etc., overlap with its reality. There is dim awareness of the other possible sequences as dreams. Together these other "realities" constitute the unconscious of the 1122 branch.

One could say that in these worlds the dream has become the reality. Here is a quantum-physical model of the Buddha's pronouncement that all "this" is a dream. It is the overlap or agreement between the different worlds that strengthens the dream, so that in any one of these worlds the worlds of thoughts (the presence of the other branches) coincide with the experienced world of the two subsystems.

Up to now each subsystem has acted as the observer of the other subsystem. Each subsystem has the same degree of consciousness and views the other subsystem as "IT" while experiencing itself as "I." The I/IT boundary arises through the interaction that, although coupling the two subsystems, also separates them.

The Quantum Physics of Agreement

To agree or not to agree, that is the real question. How can there be an apparent sensible world of law and order if every time anyone observes anything the thing quantum-jumps all over the place? The answer is simple: There are no jumps. We simply are insensitive to the changes and radical transformations of utter chaos that are continually going on around us. If we were sensitive to such changes, we would all gladly be locked up in the loony bin.

When we see anything at all, we see many, many a_i's, i.e., many possibilities, occurring simultaneously without our discerning any one of them consciously. This blurring over of distinctions inevitably produces the law of averages or large numbers (see Chapter 3). By being insensitive we are law-abiding. Or, put another way, what we call the human race is that set of sequences of self-referential processes occurring in the universe that are incapable of complete differentiation. Indeed, all consciousness may operate by the failure of self-reference to make complete differentiation. It is in this sense that consciousness, in order to exist, must be imperfect. This could be the origin of Original Sin or the Buddhist concept of suffering. Therefore, these processes appear to follow unconscious laws of order and repeatability that result in the imperfect world we see around us.

What we call agreement is the recognition that what each of us observes is limited and not the truth, the whole truth, and nothing but the truth. Each of us is a faulty observing machine. Therefore all we can do is agree through compromise. There simply is no absolute right or wrong. There is no absolute knowledge. "I say te-mah-toes you say te-may-toes, let's call the whole thing off." Or let's compromise. Thus laws arise as a result of agreed-upon compromise.

The apparent absolutes of human behavior are also due only to culturally agreed-upon norms. These norms are powerful. We are all easily convinced that the illusion provided by the continuity in time of these norms is absolute.

For example, in our culture sexual feelings are both abnormal and normal at the same time. They are normal if they occur among two consenting adult (over-18) heterosexuals in private surroundings with the doors and blinds shut. They are abnormal if any of the above is missing or altered in any way. Some alterations are "simply, snicker, abnormal." We all know they occur, but we snicker in disbelief and jealousy when we hear that these indiscretions take place. Some, however, are shockingly frightening and when they occur we lock up the offenders, wishing even to do more to "safeguard our ___." (You can fill in the blank yourself or choose any of the following: children, morals, educational institutes, libraries, movie theaters, the great outdoors, beaches, the elderly, soldiers on leave, WACS, trees, food, restaurants.)

The expected or normal value we attach to behavior changes in time. What was abnormal in 1944 (such as simply being of Japanese or German ancestry in the United States or of Jewish ancestry in Germany) is normal in 1984.

How does agreement about anything arise? The answer lies in the overlap of expectations of the observers involved. To see this, consider the following simple example: A quantum coin is being observed by an observer named Fred. Fred is a simpleminded observer. He watches only one thing during his entire life, the coin. The coin is an extremely simple object; it appears with either heads or tails. At first Fred and the coin are independent of each other. We write this as:

$$F \times C = F(ah \times H + at \times T)$$

where F is Fred's quamp and C is the coin's quamp, equal to $(ah \times H + at \times T)$, a sum of heads ($H$) with probability amplitude ah and tails (T) with probability amplitude at. But after Fred "sees" the coin, that is, interacts with the coin, he splits into two possible "Freds." F is now coupled to C and involved with it.

$$F \times C \longrightarrow ah \times Fh \times H + at \times Ft \times T$$

One Fred sees a head (Fh) with the probability amplitude ah while the other Fred sees a tail (FT) with the probability amplitude at.

At precisely this moment Jack enters the room. He watches Fred and the coin. At first, before he sees anything, he is uncoupled from the experience of Fred.

$$J \times (ah \times Fh \times H + at \times Ft \times T)$$

But now Jack observes and sees Fred and the coin.

$$J \times (ah \times Fh \times H + at \times Ft \times T) \longrightarrow ah \times Jh \times Fh \times H \\ + at \times Jt \times Ft \times T$$

Now Jack has entered Fred's world. Jack sees Fred. Jack sees the coin. Fred sees the coin. Fred sees a head (Fh) and simultaneously Jack sees the *same* head (Jh) with the same probability amplitude ah.

Or, Jack sees Fred. Jack sees the coin. Fred sees the coin. Fred sees a tail (Ft) and simultaneously Jack sees the *same* tail (Jt) with the same probability amplitude at.

We can repeat this simple observation when Judy enters the room and sees the coin, Fred, and Jack. As long as a "good observation" has occurred, each observer is in agreement.

But hold on, what about reality? So far there is agreement in probability only. When does the qwiff pop? When does Fred say, "I see heads"? The usual answer is that no one knows. Somehow one of these "branches" for heads or tails is chosen as the actual thing. Quantum physics shows how after a good observation, which means that each observer sees each coin clearly as having a head or tail showing, each participant "feels" as if he or she saw something the other participants agree that they saw.

But no one can say which branch is the real branch. Now suppose that instead of just one coin there were two.

$$F \times C \times C = F(ahh \times HH + aht \times HT + ath \times TH$$
$$+ att \times TT)$$

This corresponds to the four possible outcomes, with the probability amplitude ajk (*ahh,* etc.) standing for the coins appearing with sides j and k showing. But now suppose Fred, because of his insensitivity, cannot distinguish between *HT* and *TH.* He blurs into indistinction the middle two terms so that his branching consciousness appears as:

$$F \times C \times C \longrightarrow ahh \times Fhh \times HH +$$
$$F^{**} \times [aht \times HT + ath \times TH]$$
$$+ att \times Ftt \times TT$$

In the middle two terms, those in the brackets, [ ], Fred's mind is in two places at the same time. He has distinguished only a head/tail (the coins have different sides showing), eliminating only the possibilities that either both coins are heads or both coins are tails. He has not attempted to see which coin has which side showing. In the middle two terms, Fred is living in two universes at the same time.

If we carry this further to include three coins, four coins . . . , finally to billions upon billions of coins, we find that the middle terms in the expansion of the branching consciousness, those terms that have roughly the same number of heads and tails showing, "overcome" by sheer magnitude any of the side terms in which a sequence with more heads or more tails occurs. As long as Fred does not distinguish and only sees the blur of billions upon billions of coins (you can now change coins to atoms, and sides of coins to spin states), Fred sees only the blur and loses any sense that his observation is playing any role in the sequences of events. Fred doesn't feel coupled to the coin at all, for all that has taken place is

$$F \times C \times C \times C \times \ldots \longrightarrow$$
$$F^{***} \ldots \times [ahh \ldots tt \ldots \times HH \ldots TT \ldots + \ldots]$$
$$+ \text{ a small number of "freaky" terms with lots}$$
$$\text{more heads than tails and vice versa.}$$

All of the ". . ." terms lie together, nestled within the central brackets of life, unobserved and undifferentiated. The sequences that are "freaky" occur so infrequently as to make no odds. When Jack and Judy come on the scene there is no disagreement only so long as all fail to observe any differences in these terms. But if Judy happens actually to see a particular set of terms within the square brackets she will, by her sensitivity, not agree with Fred and Jack. She has actually entered into another universe, the parallel universe of discord.

Now we see how the observer, in failing to discriminate one distribution of coin tosses from another (as in the failure to distinguish between cases HTT, THT, TTH, the observer simply sees three tosses with only one head showing), tends to lose power and any sense of being coupled to the coin. He fails to see that he counts in the universe because after observation he still remains outside the brackets that enclosed the undifferentiated terms.

Perhaps it is a good idea to look at this once again, this time with three coins.

$$
\begin{aligned}
F \times C \times C \times C = F(&ahhh \times HHH + ahht \times HHT \\
&+ ahth \times HTH + ahtt \times HTT \\
&+ athh \times THH + atht \times THT \\
&+ atth \times TTH + attt \times TTT)
\end{aligned}
$$

This corresponds to the eight possible outcomes, with the probability amplitude *ajkl* (*ahhh, ahht,* etc.) standing for the coins appearing with sides *j, k,* and *l* showing. And now suppose Fred cannot distinguish, because of insensitivity, between HHT, HTH, and THH. (He also will fail to distinguish TTH from THT or HTT for the same reason.) As far as he is concerned, these are all the same. He blurs into indistinction these three terms so that his branching consciousness appears as

$$
\begin{aligned}
F \times C \times C \times C \longrightarrow\ &Fhhh \times ahhh \times HHH \\
&\text{term 1} \\
+\ &F^{***} \times [ahht \times HHT + \\
&\qquad ahth \times HTH + athh \times THH] \\
&\text{term 2} \\
+\ &F^{***} \times [ahtt \times HTT + atht \times THT \\
&\qquad\qquad\qquad + atth \times TTH] \\
&\text{term 3} \\
+\ &Fttt \times attt \times TTT \\
&\text{term 4}
\end{aligned}
$$

Fred's awareness is broken up into four possibilities. Terms 2 and 3 occupy his attention because these terms each consist of two like coins and one different — the more likely occurrence with three coins. Now if Fred additionally fails to see the difference between HTT and THH (he can only tell that all coins are not alike), terms 2 and 3 are joined together giving

$$
\begin{aligned}
F \times C \times C \times C \longrightarrow\ &Fhhh \times ahhh \times HHH \\
&\text{term 1} \\
+\ &F^{***} \times [ahht \times HHT + ahth \times HTH + athh \times THH \\
&\qquad + ahtt \times HTT + atht \times THT + atth \times TTH] \\
&\text{term 2} \\
+\ &Fttt \times attt \times TTT \\
&\text{term 3}
\end{aligned}
$$

257

The middle term now occupies six times as much attention as the first or the last term. When the experiment is repeated, term 2 will arise again with Fred's feeling that his observation has little to do with the outcome. His F^{***} consciousness lies outside the undifferentiated mass of mixed terms inside the brackets.

By and large, consciousness observes term 2. Consequently F^{***} appears to remain outside and independent of the processes of consciousness. (The *** simply means indifference to the result observed as long as it isn't too "freaky." HHH or TTT would appear "freaky.") Furthermore, as long as Fred has failed to observe the difference between any of the six terms within the square brackets, there simply *is* no difference. The world undifferentiated but apart from Fred remains the external and uncaring universe.

The Rules and Procedures of the Parallel Universal Consciousness Kompany, Incorporated

Looking back at our two interacting subsystems $|a>$ and $|b>$ described earlier, we could say that the system, composed of the two subsystems, undergoes a splitting interaction and primitively evolves.

$$|a>|b> \longrightarrow |a1>|b1> + \ldots \longrightarrow |a11> \ldots$$

In other words,

$|a>|b> \xrightarrow{\text{1st split}}$ terms like $|a1>|b1>$

$\xrightarrow{\text{2nd split}}$ terms like $|a12>|b12>$

$\xrightarrow{\text{3rd split}}$ terms like $|a112>|b112>$

$\xrightarrow{\text{4th split}}$ terms like $|a1121>|b1121>$

$\xrightarrow{n\text{th split}}$ terms like $|a112 \ldots >|b112 \ldots >$

With each split the number of terms is doubled: After the first split, two terms; second split, four terms; third split, eight terms, . . . ; after the nth split, there are 2^n terms.

Each split is an evolution in consciousness. All possible sequences are represented in each subsystem in this gigantic ket. There are so many that for all practical purposes the system is composed of two subsystems, both in identical states. However, no one outside the system knows what is going on. If another system like it is brought forward, it too will be described by two subsystems in identical states. However, the second system will also be just as likely to be found with its subsystems in state 2 as in state 1. After inspection of an infinite number of systems, there will be found equal numbers of systems with subsystem states of 1's and 2's. This fact makes the parallel world interpretation powerful. It leads to a natural probability interpretation without additionally assuming it.

Although no one outside the original system knows it, each subsystem possesses a knowledge of the other. For the two simple subsystems, a primitive consciousness and an unconscious arises. The sheer majority of the branches composing the equally likely distributions of states that are in resonance produces this reality. Thus, here exists a means for altering the crapshoot so that an evolution in consciousness corresponding to unlikely choices — unlikely sequences — can occur.

To understand this more fully we need to go back again to Chapter 8 (pages 199 – 202) and consider the operators that alter the kets. Looking again at Law III:

$$< I | A | A i > \; = \; a i < I | A i >$$

we see that the role of the operator (which corresponds to the previous existence of an observer seeking to measure property A of a system), A-op, $| A |$, is to produce a value $a i$ that is the characteristic of the system when it is in the state $A i$ described by ket-$A i$, $| A i >$.

I-bra-ket-$A i$, $< I | A i >$, is the quamp question: If the observed system is in the state $A i$, is it in the state I? This question is answered with a resounding yes if $A i$ and I are identical. However, it is not generally answered no if that is not the case.

The answer, then, is an infinite number of maybes. Each maybe is a number (in general a complex number — see the appendix to Chapter 8) whose relative size determines how close the answer is to being yes.

Another way to think about any quamp $< b | a >$ is to view it as a transition - probability amplitude. It contains the possibility of transformation from one state, a, to another, b.

Returning again to just one system, at this primitive level subsystem 1 can be considered to be operating on subsystem 2 in order to "realize" a value. If the interaction is repetitive instead of "splitting" as above, so that after a large number of interactions, the state

$$| a11 \ldots 1 > | b11 \ldots 1 > \; + \; | a22 \ldots 2 > | b22 \ldots 2 >$$

is produced, one says that a has performed a "good" measurement on b.

In the usual qwiff-popping or Copenhagen interpretation, a is a sufficiently complex system itself, much more complex than b. The record of repeats is usually attributed to a's memory of b. System a is said to be performing a proper, or eigenvalue, experiment on *b*. The qwiff is in a stable repetitive pattern or stationary state.

One could say in the Copenhagen interpretation that subsystem *b* has jumped into state 1 by identifying as the conscious observer *a* in memory state 1. Similarly, in the parallel world subsystem *b* has jumped into state 2 by identifying as the conscious observer *a* in memory state 2.

The prime difference between the parallel worlds interpretation and the Copenhagen one is that the latter insists, in its mechanical either/or logic, that 1 or 2 has arisen but not both. Parallel worlds theory (jokingly consisting of two parallel Copenhagens) insists that both 1 and 2 have arisen but not either.

Here one (*a*) ignores one's own presence and watches the result (*b*), witnessing a quantum-jumping world, in the case of the primitive evolutionary splitting world, or the continual repetition of a "good" value, in the case of a "good" measurement interaction. Only these two kinds of interactions — interactions that are "good" and do not induce transitions or changes and interactions that are "bad" and cause quantum jumps or transitions — are necessary to create the world, indeed the whole universe including consciousness.

So now imagine that you are in the future as the president of the Parallel Universal Consciousness Kompany, Incorporated (P.U.C.K.). Your company manufactures a pill that induces "midsummer night's dreams," you know, fantasies you are afraid to tell your spouse about. This chemical carefully alters the number and type of interactions the sleeper's nervous system undergoes. "Good" and "bad" interactions are now called repetitive and splitting interactions, respectively. Careful control of the time-release factor induces lucid dreaming in which the dreamer experiences him/herself as asleep in bed and alive, well, and conscious in a parallel dream world. Careful control alters the probabilities, so that the person "becomes" the person in the dream world and "dreams" that he/she is asleep in bed. Careful control allows the person to awaken in the dream world with only a dim awareness that there ever was anyone asleep in bed in the first place.

This fantasy is real.

RULE 1. ENTERPRISE
Virginity Break, or the Rule of Creation.
To Create, Contaminate.

(Contamination, for those who react negatively to this word, means to become what you interact with. Both "it" and "you" are thereby contaminated.) In quantum language:

$$|I>|IT> \longrightarrow |I1>|IT1> + |I2>|IT2> + \ldots \text{ (don't stop).}$$

(In whatever state "IT" is and "I" am. If "IT" is in state 5 (*IT5*), then "I" am changed by my knowledge that "I" know this, (*I5*).

RULE 2. GOODIE-GOODIE GOD

Pleasure Production, or the Rule of Hedonism.

If it Feels Good, Do It Again. Security. YES.

(Pleasure is repeat. Repeat is pleasure. Pain is pleasure repeated at a different frequency.) In quantum language:

$$|I> |IT> \longrightarrow |I1> |IT1> + |I2> |IT2> + \ldots \text{(don't stop)} \longrightarrow$$

$$|I11> |IT11> + |I22> |IT22> + \ldots \text{(don't stop)} \longrightarrow$$

$$|I111> |IT111> + |I222> |IT222> + \ldots \text{(don't stop)} \longrightarrow$$

$$|I1111> |IT1111> + |I2222> |IT2222> + \ldots \text{(please don't}$$
$$\text{stop)} \longrightarrow$$

(I feel wonderful. The world conforms to my expectations as long as "I" conform to "IT." Don't rock the boat, please. I enjoy my pain and suffering because everyone feels the same as I do.)

RULE 3. EVIL — SATAN

Tradition Breaking, or the Rule of Splitting.

If it disrupts, don't do it again. Fear. NO.

(Nothing is repeated. Everything changes suddenly. I jump into the unknown. I split into many.) In quantum language:

$$|I> |IT> \longrightarrow$$
$$\text{1st split}$$

$$|I1> |IT1> + |I2> |IT2> \longrightarrow$$
$$\text{2nd split}$$

$$|I11> |IT11> + |I12> |IT12> + |I21> |IT21>$$
$$+ |I22> |IT22> \longrightarrow$$
$$\text{3rd split}$$

$$|I111> |IT111> + |I112> |IT112> + |I121> |IT121>$$
$$+ |I122> |IT122> + |I211> |IT211>$$
$$+ |I212> |IT212> + |I221> |IT221>$$
$$+ |I222> |IT222> \longrightarrow$$
$$\text{4th split}$$

$$|I1111> |IT1111> + |I1121> |IT1121> + |I1211> |IT1211>$$
$$+ |I1221> |IT1221> + |I1112> |IT1112>$$
$$+ |I1122> |IT1122> + |I1212> |IT1212>$$
$$+ |I1222> |IT1222> + |I2111> |IT2111>$$
$$+ |I2121> |IT2121> + |I2211> |IT2211>$$
$$+ |I2221> |IT2221> + |I2112> |IT2112>$$
$$+ |I2122> |IT2122> + |I2212> |IT2212>$$

$$+ \ |I2222> |IT2222> \ \xrightarrow{\hspace{2cm}}$$

$$n\text{th split}$$

... (please stop)

(Too much change. Too much diversity. I am afraid. I can envision all possible worlds; some are too beautiful and some are too grotesque. I need to repeat once again.)

Self-Realization and God Realization in Parallel Worlds

Finally, consider the development of self-realization through the ideal laws of quantum mechanics. Where do you draw the line between psychological states of awareness and external or physical states that we can be, and are, aware of? The primary idea here is that ideal laws of reality are realized when it is appreciated how God consciousness projects itself into the separate consciousnesses that we all experience as our individual selves.

This God consciousness reality, which I called the big "I," or the $<I|I>$ quamp, is the quantum-mechanical probability amplitude of Yogananda's ultimate "self-realization."

I have chosen to approach these higher, or ideal, laws through the inductive-intuitive sense that we all possess. By observing the counterparts of ideal laws as they appear to us in the physical world, the laws of physics, and their mathematical language structures, we are able to "intuit" or inductively guess the higher or ideal laws of the universe and God.

If we ignore (as physicists did in the early discoveries of quantum physics) our acts of consciousness, we will get caught in the physical paradoxicalness of materiality. The *root cause* of this ignorance is the mind's ability to identify or become one with that which is its own projection and still maintain itself as separate from what it identifies with. Just as I feel my foot as my foot and see the star in the star's position in the sky, I am able to project my inner world onto the outer world. This experience of projection provides the double-edged blade of separateness from the world and perception of that world as it occurs in space and time. This is the life we all "sense." It is important to recognize, early in the game, our mental roles in the universe. This leads to an intuitive-inductive sense that while somehow I am not separate from that physical world, yet I feel so alone and apart from it. This double sense of aloneness/apartness and togetherness/perception is the identification *process* itself. It is also the realization of certain qualities or perceptions about the world.

At some level of this projection the very existence of the physical world is "created."

In the language of Chapter 8, we write the big I state as ket-I, $|I>$. With this "conception" we conceive of all that which is not-I. We write this as the "world state," ket-W, $|W>$. We can also write this as the not-I state, ket-$(\sim I)$, $|\sim I>$ (the $\sim$ means that which is not). Of course there is no overlap between *I* and not-*I*. Or, *I*-bracket-not *I* is nothing, not possible. I cannot be both *I* and not *I*. If I am *I,* then I am not not-*I*. To compensate for our difficulties with dealing with double negations, I will use *W* to represent that which is not *I*. *W* stands for the "world not including myself."

Thus I conceive of the state of independent and therefore unknown or unconscious kets, the $|I>|W>$, read ket-*I*-ket-*W*. These are separate unconscious ideals, yet to be realized or experienced consciously. Each can be thought of as the components of the original split, the yin-yang, the fundamental duality, the basic act of creation that began the world. *I*-bracket-*W* or *W*-bracket-*I* is nothing, no overlap. They are orthogonal to each other. Each exists in a "space" as if the other were not present.

The double flow can go $<I|<W|$ multiplies $|I>|W>$. The result depends on how the flows combine. Either we get

$$<I|I><W|W> = 1$$

or

$$<I|W><W|I> = 0$$

It's all or nothing. If it's all, and unity is perceived, it is self-perception only. The world exists and I exist, but I do not know the world and it is not conscious — it is a dead world. If it is nothing, then I know nothing of the world (*W*). It remains part of my unconscious.

To know the world there must be interaction; *I* must interact with *W*. To have this interaction there must be time. Time is the displacement through which interaction is sensed. The operation of time displacement is this interaction.

Using Law IIb of Chapter 8, both the world (*W*) and the perception of that world (*I*) are composed of infinite possibilities. Each infinity is also a unity. These unities are expressed as:

SUM (over all *i,* from $i = 1$ to $i =$ infinity) $|i><i| = 1$

and

SUM (over all *i,* from $i = 1$ to $i =$ infinity) $|Wi><Wi| = 1$

Thus ket-*I* projects into its little *i* states and ket-*W* projects into its little *Wi* states. These are written as:

$|I> =$ SUM (over all *i,* from $i = 1$ to $i =$ infinity) $|i><i|I>$

$|W> =$ SUM (over all *i,* from $i = 1$ to $i =$ infinity) $|Wi><Wi|W>$

The ket-*W* first "becomes," through projection and the law of projection, the world states, $|Wi>$. The ket-*I* "becomes" the *i*'s of the world, the observer states able to perceive these wondrous world states,

Wi. Thus I and W project into complete kets of $|i>$ and $|Wi>$. The mind does this as separate unconscious thoughts. Or we can write this: "I am this whole world yet to be realized."

Also, I-bra projects into its little i states and W-bra projects into its little Wi states. These are written as:

$$<WI = \text{SUM (over all } i, \text{ from } i = 1 \text{ to } i = \text{infinity) } <WI Wi><Wi|$$

$$<II = \text{SUM (over all } i, \text{ from } i = 1 \text{ to } i = \text{infinity) } <I|i><i|$$

This is the law of the one mind projecting. But now comes the identification process: Mind interacts with its own projections. From an intial clear mind, $|I>$, and a projected world picture, $|W>$, the state $|I>|W>$ becomes through interaction:

$$\text{SUM (over all } i, \text{ from } i = 1 \text{ to } i = \text{infinity) } |i>|Wi><Wi|W><i|I>$$

Each little i mind is coupled to its world view, Wi, and each little mind is independent of its world view at the same time. This coupling is reflected in each term in the infinite sum.

The possibility of a world view depends on the quamps Wi-bracket-W, $<Wi|W>$ and i-bracket-I, $<i|I>$. It is important to realize that big I and big W are no longer independent. I has in a sense become W. "I am this whole world" is a reality. God and the universe are one. They are entangled through time. Each little i sees its little wi, its little world. The big I is not perceptible even to itself. God is blind to God. God can only see through us, the little i's.

Since this is an infinite sum and each term in the sum can be considered a being, there is no limit to the number of beings in the universe. Now i and Wi are "one." I am no longer "I am that I am." Instead I am "that which I perceive I am." These "souls" at this point can be separate bodies or part of a single body. When we say we are making up our minds, we are composing a unity of little i's. When we say we are agreeing with each other, we are also composing a single mind, a single point of view.

11

Quantum Psychodynamics

There is a wolf in me . . . fangs pointed for tearing gashes
. . . a red tongue for raw meat . . . and a hot lapping of
blood — I keep this wolf because the wilderness gave it to
me and the wilderness will not let it go.

There is a fox in me . . . a silver-grey fox . . . I sniff and
guess . . . I pick things out of the wind and air . . . I
nose in the dark night and take sleepers and eat them
and hide the feathers . . . I circle and loop and
double-cross.

There is a hog in me . . . a snout and a belly . . . a
machinery for eating and grunting . . . a machinery
for sleeping satisfied in the sun — I got this too from
the wilderness and the wilderness will not let it go.

There is a fish in me . . . I know I came from salt-blue
watergates . . . I scurried with shoals of herring . . . I
blew waterspouts with porpoises . . . before land was
. . . before the water went down . . . before Noah . . .
before the first chapter of Genesis.

There is a baboon in me . . . clambering-clawed . . . dog-faced
. . . yamping a galoot's hunger . . . hairy under the
armpits . . . here are the hawk-eyed hankering men . . .
here are the blonde and blue-eyed women . . . here they
hide curled asleep waiting . . . ready to snarl and kill
. . . ready to sing and give milk . . . waiting — I keep
the baboon because the wilderness says so.

There is an eagle in me and a mockingbird . . .
and the eagle flies among the Rocky Mountains of my dreams
and fights among the Sierra crags of what I want . . .
and the mockingbird warbles in the early forenoon before
the dew is gone, warbles in the underbrush of my
Chattanoogas of hope, gushes over the blue Ozark foothills
of my wishes — and I got the eagle and the mockingbird
from the wilderness.

O, I got a zoo, I got a menagerie, inside my ribs, under my
bony head, under my red-valve heart — and I got
something else: it is a man-child heart, a woman-child
heart: it is a father and mother and lover: it came from

266

God-Knows-Where: it is going to God-Knows-Where — For I am
the keeper of the zoo: I say yes and no: I sing and kill
and work: I am a pal of the world: I came from the
wilderness.

Carl Sandburg
"Wilderness"
(from *Cornhuskers*)

The Relation of Space-Time to Emotion-Intellect

In this chapter I offer the theory that events taking place in the physical world are matched to events in an internal space called the *mental space*. The physical events are classified according to time, spatial location, energy, and momentum. In a similar manner the internal events can be classified as thoughts, sensations, feelings, and intuitions in the Jungian sense. Thus every physical event has a corresponding mental event. The quality of that physical event corresponds to a parallel quality of a mental event. Just as physical events exhibit complementarity such as the wave-particle duality, mental events also exhibit a corresponding complementarity as for example in the Jungian functions mentioned above. In this way the laws of quantum mechanics form a basis for a model for the new laws of mental states — for the new psychology, the new quantum psychodynamics.

If there is to be a physics of psychology, a psychophysics, classical concepts based on mechanical cause and effect will not work. Even if someone is able to show that the mind works according to classical mechanical laws (which I doubt will occur), physicists know today that classical mechanical laws work because they are based on quantum-mechanical laws. Ultimately everything is quantum mechanical, if it is physical at all.

Quantum mechanics is the mechanics of the universe. It is the most sophisticated model devised to date to deal with uncertainties, slippery elements, indirectly related characteristics, paradoxes, and therefore this exercise in quantum psychodynamics is a rigorous first step in the unveiling of the physics of the psyche.

It is also a first step in the construction of a true artificial intelligence. If the human mind and a computerized thinking apparatus are to become one and the same at some point in our evolution, it will not take place through means currently used by artificial-intelligence creators. To grasp true intelligence means grasping the elusive quantum and not grasping with simple on-off elements. The human mind contains quantum "maybe's."

Some may feel that like the quote from Democritus in the opening of the book (see page 4) my victory in logically defining the mind will result in the defeat of the intellect in that it will be cut off from the senses. Some view this, with distaste, as mechanical reductionism and a step away from the holistic-spiritual movement. However, quantum mechanics is the mechanics of the

267

human spirit. To me, a quantum physics of the psyche will not result in mechanical reductionism simply because quantum mechanics itself already avoids this mistake. A logic of the illogical is a delight that leads not to the defeat of the senses/intellect but to a resounding rejoicing. The senses become enriched as a result of the intellect and the intellect becomes smarter because of the refinement of the senses.

All experiences naturally divide themselves into complementary factors. To see how this occurs we need to return to the concepts of a qwiff, or quantum wave function, and a quamp, or quantum-mechanical probability amplitude. It turns out that a quamp is the more encompassing concept. A qwiff is, in other words, a special case of a quamp.

Briefly reviewing, a qwiff, remember, is my shorthand nomenclature for the quantum wave function. It represents a wave in space and time that, when multiplied by itself, yields the probabilities of certain events — namely, those described by the qwiff — occurring at locations in space. The quamp is a probability amplitude composed of two parts, a ket and a bra. The ket is an assumed initial condition, written ket-*a,* and the bra is an assumed final conditon written *b*-bra. Together they form a question *b*-bracket-*a,* which is interpreted as: "If *a* is true, then is *b* true?"

The connection between a qwiff and a quamp was discovered by Erwin Schrödinger in 1925 and Paul Dirac in 1926. In what was to be called the *transformation theory,* Dirac and and Pascal Jordan recognized that a qwiff was a particular kind of quamp. All that was needed to transform from one *representation* to another was to let the bra part of the quamp, $< \, |$, represent the "state" of space — the physical location of the event under question in the ket part of the quamp, $| \, >$. Thus the quamp $<x|a>$ is the probability amplitude that the state *a* is located at position *x,* or, in other words, $<x|a>$ (read *x*-bracket-*a*) is the question: If *a* is true, then does state *a* correspond to a location *x* in space?

In the above question time plays no role. If, however, I consider the reverse question — $<a|x>$ (read *a*-bracket-*x*): If the event is located at position *x,* does it correspond to state *a?* — then time enters when the two parts of the question are multiplied together as the probability (see Chapter 8 for a description of the arithmetic of bras, kets, and quamps):

$$<a|x><x|a>.$$

The entrance of time is the simultaneous marking of the event in space. It is the realization of, what I called in Chapter 1, "here." If state *a* is possible in more than one location in space "here," simultaneously, then following Law II the state *a* becomes a sum of terms, each term a duplicate of the preceding term, only with a different *x* — a different "here," corresponding to a different location in space. By taking into consideration all possible locations in space, one defines a certainty for the occurrence of the event. In the language of quantum physics, that certainty is represented by the quamp $<a|a>$, which is the probability amplitude (PROBAMP). If *a* is true, then is *a* true? The result is true or, in mathematical terms, $<a|a> = 1$.

This is generally written:

$$<a|a> = \text{SUM (over all } i, \text{ from } i = - \text{ infinity to}$$
$$i = + \text{ infinity) of } <a|xi><xi|a>$$
$$= 1$$

Furthermore, if all these terms are equal in magnitude — for example, if

$$<a|x1><x1|a> = <a|x2><x2|a> = \ldots$$

then the quamp $<xi|a>$ represents a simple qwiff, that of a particle with a well-defined momentum moving from minus infinity to plus infinity in the *x* direction (along one dimension of space). By changing the form $<x|a>$, one changes the shape of the qwiff in the same manner that one mathematical function describes a wave moving as a plane perpendicular to itself and another mathematical function describes a wave moving as an expanding sphere or circle.

If, on the other hand, *x* is not a location in physical space but is a location in mental space (and I choose these words as carefully as I can), then the quamp can be said to describe a state of mental space.

The idea of a mental space was first formulated by Sigmund Freud when he described the space of the unconscious. Structures such as the id, the preconscious, and the ego can be said to exist in this mental space.

The "space of the psyche," states of awareness, altered states of consciousness, and other mental states are correspondingly describable in the same mathematical framework as is used to describe physical states, namely, as states in Hilbert space — the abstract mathematical space used in physics that is composed of an infinite number of dimensions. (Each dimension represents an ideal situation or state of possible existence for a system.)

In other words, physical space and mental space are spaces within the more general Hilbert space just as a chair and a sofa lie within the space of a room.

Thus, for example, a state *a* (described as a ket, $|a>$) is realizable as both a physical event in space at *x*, $<x|a>$, and as a body sensation, *y*, $<y|a>$, in the mental space.

Furthermore, just as space and time are united in a space-time map of physical events, emotion and intellect can be considered as similarly united in an emotion-intellect map of mental space. I am therefore suggesting that a state that corresponds to a location in space also corresponds to a sensation in the mind or mental space. Similarly, a state that corresponds to an event in time, having a temporal beginning and end, also corresponds to a thought in mental space.

With this realization many human problems such as the difficulties in the resolution of personal conflicts, emotional disorders, and the seemingly inescapable urge to to self-destruction may be ameliorated. If this hypothesis is true, then the impossibility of obtaining certainty on the physical plane of existence is exactly matched by a similar uncertainty in human intercourse.

For example, according to the principle of complementarity in quantum physics, it is impossible to state any set of conditions that will reproduce a dynamics of certainty about the simultaneous location and momentum of any physical system. Similarly, there should be a parallel complementarity in our mental worlds.

I theorize that these physical and mental complementarities are as shown separately in the Figures 59 and 60.

COMPLEMENTARITY OF PHYSICAL STATES

Figure 59

CORRESPONDING COMPLEMENTARITY OF MENTAL STATES

Figure 60

Energy and momentum are complementary descriptors for spatial location and time of occurrence of any event. In the wave-particle duality the wave is described by its energy and momentum (or wavelength) while the particle is described by its location in space and time of interaction. Either description scheme, energy-momentum or space-time, is incomplete because one without

270

the other cannot lead to a prediction about the future, that is, give a cause/effect relationship.

Both are needed to make predictions. That is why these two descriptors are complementary to each other. Quantum physics has shown that it is not possible to have both descriptors operating simultaneously.

In a similar manner feeling-intuition and thought-sensation can be considered as complementary descriptors for mental or psychic events. The "wave of emotions" that overcomes one is closer to the truth than we may have suspected. Feelings and intuition are the wave complement to the thought/sensation precision of the intellect. Quantum physics offers a model to explain the duality, the split-brain functioning, and the conflict we all feel when faced with having to think clearly when a situation is emotionally charged.

The key idea here is the concept of "state." In classical mechanics a state meant a simple gathering of data. For example, the state of motion of an object was the specification of its mass, position, location, energy, and momentum at a precise time. In quantum mechanics we cannot specify a state this way. Instead, a state becomes totally abstract, capable of being realized by one complementary descriptor set (energy-momentum, position-time) or the other. For example, a quantum object in a given state can be said to have either a spatial location or a momentum distribution. Either description would be a complete description of the state, but would lack predictability.

The state would be complete, but the description of the state would, because of its lack of complete predictability, be incomplete. In a similar manner all states can be represented by mental descriptors that also lack in predictability. A similar complementarity operates in mental space. Every state that is "realized" has both physical and mental descriptors. This is the new idea presented here. Both the physical and the mental representations of the state are simulacra (images). This division of representation of a state is the mind/matter split. Furthermore, just as events are complementary in the physical plane:

> *a state described by a well-defined energy cannot be described as having a beginning and/or an ending time;*
> *a state described as having a beginning and/or an ending time cannot be described as having a well-defined energy;*
> *a state described by a well-defined momentum cannot be described as having a beginning and/or an ending spatial location;*
> *a state described as having a beginning and/or an ending spatial location cannot be described as having a well-defined momentum.*

These same states exhibit a similar complementarity in the mental plane:

> *a state described by well-defined feelings cannot be described as a well-expressed thought;*
> *a state described as a well-expressed thought cannot be described by*

271

> *well-defined feelings;*
>
> *a state described as a good intuition cannot be described as having a*
> *well-located bodily sensation;*
>
> *a state described as having a clear bodily sensation or location cannot be*
> *described with good intuition.*

It will be helpful to think of the two worlds, mental and physical, as parallel planes. Their overlap is shown in Figure 61.

OVERLAPPING OR PARALLEL COMPLEMENTARY STATES

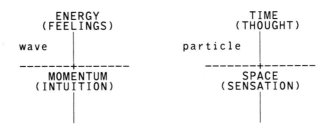

Figure 61

This description of mental experience as parallel to physical experience I have termed *quantum psychodynamics.* Just as quantum electrodynamics describes the dynamics of electron and photon interactions, and quantum chromodynamics describes the interactions of quarks and gluons, there is a quantum dynamics that describes the psyche.

The problem here is that we don't have a particle of the psyche that is identifiable. In fact, the concept of a particle is really limited to space-time descriptors. This has led me to the concept of the *morpheme.* In linguistics, a morpheme is the smallest unit of speech that meaningfully distinguishes one utterance or word from another. The *m* in *mother* and the *k* in *kiss* are English morphemes.

The ancient Hebrews believed that the universe was created by God's words and that sounds form the physical world. It is clear from our Western industrialized state that words and symbols make up our creation, the Western world. Since physics itself is impossible without words, it appears logical that word units, or morphemes, play a similar role in the construction of the psyche as electrons play in the construction of the physical world.

272

Sanity and the Nondivision of Thought-Sensation and Feeling-Intuition

In the following we will look at how the two planes, the mental and the physical, interact or intersect. But first it is necessary to understand how consciousness divides itself into the mental (mind) plane and the physical (body) plane of what we call existence.

From the quantum-jumping point of view, this division is caused by the actions of the qwiff pop. In the parallel worlds hypothesis it is more subtle. According to it, there really is no division taking place. This physical existence is no more important than any other physical existence that appears to this one only as "mind" or in dreams.

To "you" in the other existences, "this existence" is a dream or just a passing thought lasting perhaps a second. (Remember the "flea-trip" described in Chapter 1. Each second of "flea-time" corresponded to around seven days of earth time by the end of the flea's fourteenth year in space.)

Thus the quantum-mechanical laws described in Chapter 8 are applicable to our minds as well as our bodies, because our minds are parallel natural worlds to this one. Thus if quantum physics applies here, it must also apply there. And, conversely, we can learn about our minds by inquiring into our processes of observation. As in any good detective novel, the clues to how and why we think the ways we do lie "out there" in the physical world.

Laws of behavior are traceable to the new physical laws. The distaste most of us feel about mechanical behavior or nonrational mysticism is the chief reason that attempts to link physics and psychology have met with difficulty. We intuitively know we are more than machines. Agreed, but what is missing? I suggest that we look again at human behavior from the vantage point of the "new physics." Let's examine the quantum-physical laws as they pertain to human behavior.

The principle of uncertainty, for example, is such a law. Because it is stated in neoclassical, prequantum language, the mental counterpoint, the mental actions of the observer, are left out. In the classical-mechanical physical plane, Newton's laws should suffice to explain matter — there is no reason for uncertainty. But, there *is* uncertainty and it is fundamental and necessary to all existence. This is difficult to understand because the parallel or mental component of that uncertainty — the reaction of the "mentality" (thoughts, feelings, sensations, intuitions) of the observer of the tiny atom of matter under scrutiny — is left out.

In the qwiff-pop model the observer pops the qwiff by his act of observation. This pop is influenced by his thoughts, feelings, sensations, and intuitions. The thought of the observer plays an essential role in choosing what can be measured. The intuition of the observer sees more than just data as a result of the measurement. Similar feelings arise in the observer when certain properties of the object are seen to recur.

273

In the parallel world model the observer's interaction with the observed branches him onto different parallel worlds. With each thought there is a world. With each sensation there is a world. The sequences of thoughts and sensations, when examined in hindsight, exhibit a systematic causality with an uncertainty principle in operation. This uncertainty arises from the branching of the qwiff that is caused by the interaction. Each branch fails to "see" its common branch point.

Now why is there any division between the mental world and the physical world? The answer according to my view of the parallel world hypothesis is *phase harmony.*

The mental worlds are, according to the parallel worlds concept, actually occurring now. What you think here is happening there. The forms of physical reality we experience are not unique. They are formed from harmonies of other parallel realities. This physical existence is just one of many mental existences. It would be better to think of consciousness as capable of being represented on many reality planes. This existence is composed of certain planes in resonance (see Chapter 9). And these in turn divide into physical existence and mental existence for each separate plane. On our plane of existence the mental existence comprises not only our emotion-intellects but that of all of the rest of the universal layers of consciousness — in other words, the thoughts, feelings, sensations, and intuitions of all sentient beings.

Once it is recognized that these parallel layers influence us here and now, certain archaic concepts of psychology are bound to fall by the wayside. For example, the concept of multiple personalities is due for a fall. All of us are multiple. The ability of certain individuals to manifest entirely different personalities was shocking in the fifties and sixties, but today research indicates a far more accepting approach.

So we are left with one basic question: Why? For what purpose is the universe constructed in such a bizarre manner? Wouldn't it have been much easier to have just a simple physical universe without consciousness entering or parallel worlds splitting? Yes, of course it would have been far simpler. However, its complexity tells us something. As any good magician will tell you, magic tricks are "put over" by making the audience focus on unessential details. Perhaps consciousness has invented all this craziness to distract it from discovering its own essential nature of unity. If so, then the discoveries of modern physics and psychology will lead to the uncovering of the Magician at work in all of these parallel universes. This discovery arises as the force of evolution, the movement called time.

Evolution from Parallel Worlds

The "movement" of time is expressed as evolution. The direction of evolution may be described as making matter conscious. From the parallel worlds hypothesis, this means bringing more worlds into harmony through inter-

274

actions that lead to the greater acceptance of physical and mental differences. This is its higher purpose, or its higher will.

Without its division into mental and physical planes, reality would appear even more bizarre than it does. Physical or material laws are concrete for a purpose: to break the Gordian knot of paradox that must pertain or exist "side by side" with logical reason in all the parallel or psychic levels. How this knot is broken depends on how interactions are seen from the temporal point of view.

A typical example of how parallel worlds enter human evolution is exhibited by the growth of a child. Both the child and the adult who grew from that child "live" on parallel mental levels. These levels exist side by side but are experienced as past and present by the adult and as present and future by the child. Actually both are alive in parallel universes that do not overlap or extend into each other. This is only paradoxical, however, to those who insist on one materiality — on the physical plane as the sole basis for all thought and feelings and all existence.

The totality of all parallel worlds is of great simplicity and beauty provided one is willing to give up those confining notions of the absolutist viewpoint prevailing as materiality. At the parallel worlds level of awareness certain fundamental rules at first seem contradictory. Thus, for example, consider a childish man, one who never seems to "grow up" and face responsibility. This individual is both a man and a baby at the same time.

In one world that persona manifests in the physical plane as a man. In another, parallel world he manifests as a baby. In each physical world the other persona is mental. Thus the baby sees how he is to grow to be a man, while the man exhibits his "babiness" by failing to release his memory of that other level of physical existence.

In this way both baby and man suffer. The man suffers because the man does not lead but follows the baby. The baby suffers because he does not receive nurturing from the man who will be his future. Only the man is capable of leading the baby — the man must let go of his babyhood so that the baby will receive his natural inheritance in becoming a full man.

It is only through the realization of present consciousness brought forward by enriched interactions, by appreciation of the interaction among differences in the universe of discord, that we can release our past to let it grow and change. Thus the true cause of suffering on the material plane is our unwillingness to let go of our babyhood — the babyhood of humankind, the early, reptilian adventure we call our origins. We adhere to it. The proof is the existence of reptiles at this level of being. These reptiles are both physical and mental; they are "out there" and they are in our heads. Our reptilian brain is unfortunately still running the show. Survival comes first.

The physical reptiles are there to remind us that we must let go so that they can evolve into us. All physical suffering is due to our unwillingness to let go of our prehistoric parallel world of wars, death, and violence — all necessary when humankind first began.

At the psychic level our "person" is both baby, reptile, snake, monkey, eagle, and adult at the same time. All of these psychic influences project into all of the material planes of existence that manifest. Thus the "adult" is on one universal level an ape. That ape "remembers" dimly that it has a higher manifestation. That ape is psychically aware of its humanity, but it is not really human. We say, in the language of Darwin, that the ape will evolve into a man.

The man, who is at an obviously higher level of consciousness than the ape, is also aware of his apeness. He remembers his lower manifestation. He is more aware of his apeness than the ape is aware of his humanity. Both ape and man are equally projected from consciousness. When the man lets go of his apeness, he can evolve. When the man holds on to his apeness, *both* the man and the ape cease to evolve.

The true cause of evolution cannot be just of this physical world; it is not simply due to materiality. Its true cause comes from the totality of all the parallel worlds in which we exist in all possible forms. Viewed from this world, these other worlds project into the material world as a result of our interactions with each other and of the daily events of our lives. The richer these interactions, the more we evolve as higher conscious beings.

Furthermore, this projection arises not as repeatable experiences lending stability and inevitability to this world, but as helter-skelter disruptions of our psyches and the splitting of the universe into revolutionary fragments. This is the root cause of our fear, to let go of the past, of the parallel universe in which man is a reptile.

There are many who may scoff at this. After all, what does quantum physics, which governs the behavior of atoms, molecules, and subatomic particles, have to do with the evolution of species? The answer is that molecules also evolve. The evidence for this is cited in a recent book entitled *The Origins of Life: Evolution as Creation.* The author, Hoimar von Ditfurth, offers some interesting hypotheses about evolution and the universe. Ditfurth sheds some light on the controversial debate between creationists and evolutionists.

Although he puts down religious fundamentalists and tends to be scientific in his argument, he thinks little of the scientist's materialistic explanation of the source and creation of life. He skillfully shows how creation without evolution is nonsense and shows that, conversely, evolution cannot be explained without creation. Ditfurth thinks that evolution is the Creator's way of governing the world disguised from humankind's limited perspective. But it was Ditfurth's explanation of molecular evolution that especially interested me. Just the idea that molecules can evolve sends shivers up my spine. There is in existence an enzyme, a molecular fossil, so to speak, responsible for the most basic function of any living cell, intercellular oxidation (the burning of food in the cell to produce energy). This is a very old enzyme since it carries out today the same function it did when early life began.

This enzyme, Cytochrome C, "a molecular string composed of 104 amino acids," evolved at the same rate as the species carrying that enzyme, as indicated by the 500-million-year genealogy reconstructed by means of fossil discoveries. Each mutation in that molecular chain can be matched to the appearance of the species.

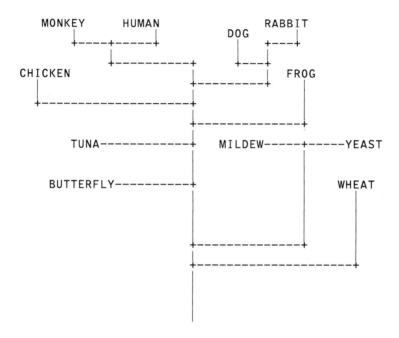

Figure 62
The genealogy derived from the comparative molecular structure of Cytochrome C. It compares identically to the genealogy reconstructed by means of fossil discoveries.

By comparing the differences in the Cytochrome C molecular strings from human, monkey, dog, rabbit, chicken, frog, tuna, butterfly, wheat, mildew, and yeast cells, Ditfurth skillfully shows that these differences match fossil discoveries of the species themselves; each species appeared when its Cytochrome C made a change in its molecular structure. In other words, an evolutionary jump.

In considering how the mind came into the world, Ditfurth resolves that "the brain does not produce mind, mind emerges in our consciousness by means of that organ." He concludes that "evolution — supposedly so hostile to religion — has shown us that reality doesn't end where our experience stops . . . evolution compels us to recognize an 'Immanent transcendence' that immeasurably surpasses our present cognitive horizon."

How does the parallel worlds hypothesis explain the evolution of the species? First, consider the classical explanation. The idea is that nature produces copious amounts of genetic material, a gene pool. This gene pool contains for the most part the genetic chain molecules that ensure the survival of the species to which that gene pool belongs. Through random genetic mutations produced by random events certain strains are created that have no chance of surviving in the environment in which they find themselves. For example, during the era in which animals lived in the sea, warm-blooded mutants — animals capable of maintaining their own body temperature, which differed from the ambient seawater — would die out. Albinos would be produced that also would not survive in the strong light conditions found in Africa. Yet these mutations continue to appear from time to time as purely random events.

It is as if nature continually was creating life forms that can survive if the proper environment existed for them but would die out unless the environment changed. Ditfurth cites the experiments of Lederberg using petri dishes of bacteria that became resistant to streptomycin. Lederberg claims that most bacteria are not resistant to the antibiotic. When placed in the petri dish nearly all of the bacteria die out, but those that survive, according to the classical evolution theory, are already resistant to it. These in turn normally die out in the natural environment since streptomycin is not present.

By separating two identically bred petri dishes of bacteria and putting into one of the dishes the streptomycin pathogen, thus killing off nearly all the bacteria, Lederberg found in one tiny corner of the dish survivors resistant to the change. In the other dish he found in the identically located corner bacteria also able to resist the pathogen. The idea was that since the two dishes were identical replicas of each other, the pathogen-resistant strain already existed in both dishes. This would mean that nature had already produced the resistant strain just in case she needed it later on when the environment might change.

But there is an alternative explanation, namely, the EPR effect (see Chapter 1). According to EPR, two systems that have had a past interaction will be correlated so that, even though they are no longer in interaction, a measurement performed on one of the systems creates instantaneous effects on the other. In the parallel worlds scheme there is no paradox to this since each possibility for the paired systems exists in separate worlds.

The two dishes were separated after undergoing identical growth patterns. This ensured that whatever genetic chain molecule was created in one dish at one site would be reproduced at a like position in the other dish, showing that the dishes were correlated — had a common origin just like the typical EPR correlation. When the experimenter changed the conditions in one of the dishes, he produced an instantaneous spacelike noncausal, EPR-linking change in the other dish.

Thus evolution would occur in all species previously correlated with each other. It is not that the bacterial strain already contained the resistant molecular arrangement. It was created when the environment "selected" that probability amplitude and turned it into an actuality. Those species evolved that had enriching and parallel-world-splitting interactions. Those that repeated the same processes over and over again did not. Later in this chapter (see pages 289 and following) we will return to the idea of evolution as a transformation of feelings into thought and sensations.

Psycho-Physical Parallel Planes

In the remainder of this chapter I wish to show in more detail how consciousness naturally splits into the physical and the psychic planes that constitute the network of conscious experiences. To do this we need to look at what I call the psycho-physical parallels. These in turn depend on certain basic concepts in physics. The primary idea on which physics depends is that of location in space and occurrence in time. But what is located and what occurs? Usually when a physicist refers to a location, he/she means an event. This event has associated with it the detection of a particle having mass. Usually this detection is possible because the particle has undergone some form of energy transformation. I shall use the rather awkward term *mass/event* to remind us that mass is an experience associated with being somewhere, at a spatial location and occurring at some time.

Connected with this concept are energy and momentum. Energy release only results when the mass/event changes. A change associated with the movement of the mass/event in space is called momentum. In quantum physics it is not possible to predict with certainty both the momentum change associated with a mass/event change and its locations in space. Similarly it is not possible to specify with certainty both the energy release and the moment of that release.

In what follows I suggest that these changes and their associated uncertainties are accompanied by parallel changes in our psyches, thus in both planes of our existence. The connection between the psychic plane and the physical plane is provided by comparing the parallel complementarities of quantum mechanics as shown in Figure 61. The sensation of an event on the psychic plane is accompanied by the appearance of a mass/event in physical space. This psychic sensation appears as a primitive morpheme or prethought. Our words are composed of sequences of morphemes. Our thoughts result. Similarly, a thought that occurs on the psychic plane is accompanied by a sequence of mass/events in the physical plane, as, for example, in our brains.

Just as a sequence of physical events cannot be predicted because of the uncertainty principle, a sequence of sensations, or prethoughts, cannot be completely determined on the psychic plane.

Elementary sensations, or prethoughts, accompanying physical events give rise to our sense of time. Our ability to speak and write, to form words, undoubtedly follows a similar format.

As I mentioned in Chapter 3, for several of my teen years I suffered from stammering, the inability to create a word from a struggling prethought. I attribute this now to a psychical uncertainty principle. As long as I held the "position" in my vocal apparatus that I must say hello, there was no movement or flow of words. By learning to relax through deep-breathing exercises and other meditation techniques, I eventually "forgot" the words and spoke without impediment.

Any learning is difficult at first because the positions of the events that constitute the pattern of that learning must be memorized by rote. After a while the learner forgets the positions and just skis down the hill or hits the ball. Indeed attempts to remember the positions usually produce the wrong results: "Strike three, you're out!"

In other words, mass/events in the physical plane we call our bodily reality are accompanied by simultaneous occurrences of sensation sequences, or prethoughts, in the psychic plane.

Just as matter transforms into energy, morphemes also undergo a transformation into feeling. For matter to move, it must continue to transform itself through space. Its content of matter, called its rest mass, changes when matter gives off energy. Even if it does not radiate away any energy, its mass will change if it simply increases its speed. The energy required to make it move faster will all go into increasing its mass.

That is what $E = mc^2$ means. By slowing the mass and thereby extracting its energy we are extracting its mass. This is just as true as if we were extracting a hunk of metal from the ground.

In a similar manner a morpheme appears as a sensation that transforms into feeling — a parallel consequence of matter appearing in space and transforming into energy inside the nervous system. Matter that receives that energy must, according to $E = mc^2$, increase its mass. Thus matter in us changes constantly. These transformations are felt as a large part of our life force. Just as matter is not fixed, we do not have a fixed weight or a fixed amount of matter within us.

If we could use extremely high-speed photography to watch our bodies (we would need high-frequency light well beyond the optical range of our eyes to do this), we would see that energy was constantly annihilating itself and creating nodules of matter as it did so. We would see nodules of matter disappear and give birth to bursts of energy — pure light — as high-frequency photons. It would appear to us like a macabre dance of great and terrifying beauty.

Similarly, the morpheme events that constitute the network of our sensations are felt as body sensations because transformations of feelings occur at the psychic level. The matter-into-energy transformation in the body is accompanied by morpheme sensation transforming into feelings. The parallel streams flow like this:

matter in space $\longrightarrow$ energy $\longrightarrow$ matter in space $\longrightarrow$ ...
morpheme sensation $\longrightarrow$ feelings $\longrightarrow$ morpheme sensation $\longrightarrow$ ...

The outer, or physical, form is called matter/energy transformation. The inner, or psychical, form is called sensation/feelings transformation. These constitute the mental universe. Together they make up the whole universe. Thus sensation needs energy (which are feelings) with which to focus — to provide the stable boundary conditions — which it does by creating nodules of matter by which a morpheme sensation can reflect on itself and thereby "know itself." Morpheme sensation is primal to the transformation of matter. Matter thus changes through the action of morpheme sensation.

What occurs to us when matter is observed to move in space? To grasp movement several morpheme sensations must be involved. A certain consistency of experience, a repeat of a sequence, like the vibration of a musical note, is "felt." The change in the movement of matter in space gives rise to feelings. These feelings are an evaluative function. For example, to experience elementary feelings about body weight, take an object in your hand right now. Even though you are not thinking the thought that this object has weight, you have feelings about it. These feelings are more than sensation alone can provide. By taking another object in your other hand, you can feel that one is heavier. Again, an evaluative function is occurring.

This evaluation acts in the psychic plane. The transformation of matter into energy on the physical plane has a correspondence on the psychic plane. Morpheme sensations transform into the movement of sensation, or feelings.

When the momentum of a mass/event is measured in quantum physics, one is usually referring to its wave property or its wavelength. The momentum of the mass/event is inferred from the formula

$$p = h/L$$

where p is the momentum, h is Planck's constant, and L is the wavelength of the mass/event. To determine this wavelength one must not attempt to measure the location of the mass/event. Attempts to do this spoil the wavelength determination so that no momentum measurement is possible. This again is the uncertainty principle in operation.

In a similar manner the intuition of the mind is tapped when one no longer "pays attention" to one's body sensations. Meditation evokes this state of intuition. Notice that this is not a state of thinking — there are no words used. Thinking gets in the way of intuition in the same manner that a mass/event's location measurement inhibits the momentum measurement. However, just as

281

all measurements of momentum eventually are recorded and turned into space-time "language," into position measurements on a recording device, intuitions and feelings are turned into thoughts through the interactions of morphemes.

In the next two sections we will look at the mathematics of the transformation of thoughts and sensations into feelings and intuitions.

The Mathematics of the Transformation of Intuitions into Sensations and Back Again

In order to grasp how sensations and thoughts manifest as feelings or intuitions, it is necessary to look at how mass/events appear as having energy and momentum. The physics of the formation of mass/events as exhibiting energy and/or momentum is governed by the mathematics of transformation functions.

One of the most powerful mathematical techniques invented is the concept of mathematical transformation. Transformation is a key concept in the "new physics." For example, the metamorphosis of a wave into a particle is represented by a mathematical transformation.

The idea of mathematical transformation was discovered in the eighteenth century by Joseph Fourier, who realized that any mathematical function that is defined over a region of space can be represented by a sum of sinusoidal, or wave, functions. These sinusoidal functions turned out to be just what was needed in quantum physics to transform a wave into a particle and back again.

I propose that similar transformations occur in the operations of the human nervous system. To understand this better, a little acquaintance with functions in mathematics is necessary.

A mathematical function is given by an algebraic equation such as

$$y = x$$

or

$$y = \sin(x)$$

It defines a relationship between what are called the dependent variable, y, and the independent variable, x. Thus if x is a given number, y is a correspondingly given number, $y(x)$ (read *"y of x"*).

A sinusoidal, or wave, function is one that repeats itself in a wavelike form called the *sine wave*. In other words, as x is allowed to change, $y(x)$ undergoes oscillations such that $y(x)$ takes on the same value at periodically located positions in space.

In quantum physics the act of observation that locates an object in space is represented by a peculiar mathematical function called by its inventor, Paul Dirac, the *delta function*. Dirac's delta function is strange. It

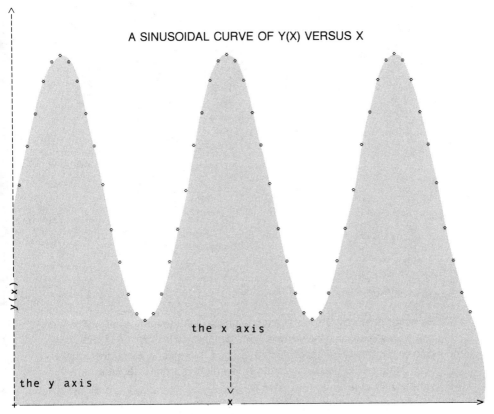

Figure 63

has the value zero everywhere except at one point, and there it has the value infinity. This weird function can be also represented by a sum of sinusoidal functions. Its representation as such a sum of sine waves is important in describing how a mass/event "propagates" in time and through space.

Dirac, remember, is the same physicist who gave us the bras and kets of quantum probability functions. As you may guess, Dirac's delta function fits perfectly into his notation. It is what we have previously written and called a quamp, namely:

$$<x'|x>$$

and corresponds to the question: "If the event is located at the point in space x, then is the event located at the point in space x'?" Of course the answer is no, or zero, if x' is not equal to x. But if x' is x, then the answer is infinity.

This may strike some readers as quite strange when it is realized that $<x'|x>$ is a probability amplitude for x being equal to x'. Dirac's function

283

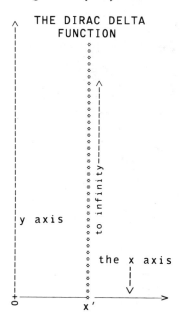

Figure 64

is never physically realized. What is physical is the integral of $<x'|x>$ over all values of x including x'. The process of integration is akin to measuring the area under the curve of $y(x)$ versus x. Since the width of the delta function curve is zero but its height is infinite, its area turns out to be unity.

In a similar manner it is possible to construct a momentum delta function quamp with the same kind of question. However, momentum is not located in physical space. One thinks of it instead as existing in something called *momentum-space.* The momentum delta function is written simply as

$$<p'|p>$$

and corresponds to the question: If the event has momentum p, then does the event have momentum p'? Again the answer is no or zero if p' is not equal to p. But if p' is p, then the answer is infinity.

As Fourier noted in general and Dirac noted in particular, it is possible to transform from one question, from one quamp, to another. It is also possible to think of this transformation as occurring from one space to another. In other words, there is a transformation from $<p'|p>$ to $<x'|x>$.

To accomplish this transformation, a quamp transformation function is needed. It is called the *Fourier transformation function* or, briefly, the Fourier transform. Fourier discovered that this function is a sine wave. It is a mathematical function that exhibits characteristics associated with both momentum and physical space. As recognized by ancient metaphysical thought, this function has a vibrational form when seen in physical space. It is represented in Dirac's quamp language as

$$<x|p>$$

and since it is a wave in space that constantly repeats itself, it has a wavelength L. Thus, according to quantum physics, such a wave represents a particle with a well-defined momentum p, given by the formula

$$p = h/L$$

(see page 279) but without a well-defined position. Thus the transformation function is also the quantum wave function for a particle with well-defined momentum. Whenever a particle is in such a state, it is called a wave. Thus the p in the ket-p part of the above quamp is the same p that is used in the formula $p = h/L$. The transformation function $<x|p>$ is also a question. The state of the transformation of the mass/event from a wave with momentum p to a particle with position x is represented by the question: If the event has momentum p, then is it located at position $x?$ — a state of known momentum but unknown position.

One could think of x and p as if they were in two different worlds, the positional world of x and the momentum or wave world of p. The transformation function $<x|p>$ represents the overlap of the momentum or wave world of a mass/event as it "bleeds" through into the positional world. Just as one picture is worth a thousand words, one momentum state is made of an infinitely large number of possible states: If you know the momentum of the event, you do not know its position.

By turning the quamp around so that it becomes

$$<p|x>$$

we have a parallel question: If the event is located at position x, does it have momentum $p?$ Here the quamp represents a mass/event "bleeding" through from position space into the abstract momentum space. Here one position is worth an infinite number of possible momentum states.

The mathematics of this transformation is nothing more than Law II of Chapter 8 carried through two times — that is, one must do two sums. First, one holds a particular "final" fixed momentum and adds up all possible initial momenta. Then one adds the results with different final momenta. The transformation from momentum to position is written

$$<x'|x> = \text{SUM (over all } p \text{ and } p') \text{ of}$$
$$<x'|p'> <p'|p> <p|x>$$

There are three questions put together in the above, with the middle question being the momentum delta function undergoing transformation from initial momentum *p* to final momentum *p'*. Starting on the right, the questions are: If the event has position *x,* does it have momentum *p?* And if it has momentum *p,* does it have momentum *p'?* And if it has momentum *p',* does it have position *x'?*" Lots of ifs.

The "bookend" quamps are the transformation functions. Think of this as an adventure. You start out at location *x* and jump in consciousness to momentum space landing at "anywave," *p.* Here you are flying with no sense of where you are. You then jump to another momentum wave, *p',* in the same space. Next you jump back from momentum space to location *x'.* You do this endlessly with all possible *p* and *p'* waves.

This means that the only way to have certainty about the location of the event in space is to have all possible momentum states; or, in other words, the location of an event in space is an overlap of an infinite number of parallel momentum universes, in each and every one of which the momentum of the event is well defined.

It is also possible to transform back from position or physical space to momentum space following a similar procedure. The transformation from position to momentum is written:

$$<p'|p> = \text{SUM (over all } x \text{ and } x') \text{ of}$$
$$<p'|x> <x'|x> <x|p>$$

This sum shows that the only way to have certainty about the momentum of the event is to have all possible position states; or, in other words, the momentum of an event is an overlap of an infinite number of parallel position universes, in each and every one of which the position of the event is well defined.

Since we experience all events in terms of physical space (we even think of internal events using the language of physical space, as when we say "I see" to mean "I understand," momentum space is not directly experienced. What is experienced as momentum is a relationship between events having locations in space. This relationship is observed by looking at the transformation from momentum to position.

Although we usually don't think about transformation functions that take us from physical space to wave or momentum space, there are many examples to be found in everyday life. A typical example of this transformation experience is found in the ordinary lens or pinhole camera. Photography works because the pinhole is the position Fourier transform of the momentum information contained in the light waves passing through it. The light wave amplitudes from many different locations in space (the object being photographed) come together at the pinhole or at each point of the camera lens. This is the transformation from wave to particle, to a single point in space. Thus the pinhole quamp $<x'|x>$ is according to Law II a sum:

$$<x'|x> = \text{SUM (over } p \text{ states) of } <x'|p> <p|x>$$

where each $<x'|p>$ is a light wave. These light waves are superposed on each other. The pinhole point contains a sum of waves. Later these waves spread out as they seek the film in back of the camera. When the waves hit the film, they "collapse" again to points on the film, thus reconstructing the image of the photographed object. This whole movement is a series of transformations from wave to particle to wave and back again to particle.

The mathematics of the transformation of intuition into sensation and back again follows a similar format, only this time we are dealing with psychic space, not physical space. Thus s represents a primary morpheme sensation, or briefly a sensation in an abstract nonphysical internal sensation space. Just as there is a physical space, there is a sensational space. Events in physical space are marked by their spatial location and their temporal occurrence. As we will see in the next section, a change in energy marks, or creates, the measure of time. In a similar way changes in feelings can mark or create thought. Changes of feelings appear as thoughts. Just as time distinguishes a past from a future in a memory record, thought distinguishes one sensation event from another. And just as momentum is conjugate to position (you can't know one if you know the other), intuition is conjugate to sensation.

Therefore, to transform from sensation into intuition and back again, a similar Fourier transform is used. It is represented as

$$<s|q>$$

and it is also a morpheme wave that has a constant "wavelength" in sensational space. And consequently the q in the ket-q part of the above quamp has the same interpretation as the p in the formula $p = h/L$. It is not obvious to what this wave corresponds in mental life. It could be the wave described in Chapter 8 (see page 211) as the life wave of attention. It undoubtedly has much to do with sound, but not necessarily physical sound; although physical sound is a manifestation of this inner psychic sound or sound energy.

Thus it, too, is a question. In fact, you exerience the abstract sensation-intuition transformation of this inner sound when you are actually thinking of any question. The state of the transformational question — If the event has intuition q, then is it a sensation s? — is a state of known intuition but unknown sensation. This means that it is the quamp representing the intuition of an event as it materializes into sensational space.

By turning the quamp around so that it becomes

$$<q|s>$$

we have a parallel question: If the event is a sensation s, does it have intuition

287

q? Here the quamp represents an event materializing in abstract intuition space.

> The mathematics of this transformation is again Law II of Chapter 8. The transformation from intuition to sensation is written
>
> $$<s'|s> = \text{SUM (over all } q) \text{ of } <s'|q> <q|s>$$
>
> I have simplified the formula by using the fact that $<q'|q>$ is zero unless q' is q. This is how the $<q'|q>$ question is transformed into the $<s'|s>$.
>
> This means that the only way to have certainty about the sensation of an event is to have all possible intuitional states; or, in other words, the sensation of an event is an overlap of an infinite number of parallel intuition universes, in each and every one of which the intuitions of the event are well defined.
>
> The transformation from sensation to intuition is similarly written (see pages 283–85):
>
> $$<q'|q> = \text{SUM (over all } s') \text{ of } <q'|s'|q>$$

This means that the only way to have certainty about the intuition of the event is to have a superposition of all possible sensation states; or, in other words, the intuition of an event is an overlap of an infinite number of parallel sensation universes, in each and every one of which the sensations of the event are well defined.

A well-defined sensation is one clearly identified as actually taking place within the nervous system, such as the sensations of heat, light, sound, smell, or taste. It is the sensation space marking of a physical space event. The "cause" of the event need not be physical, however. Pseudosensations such as painful phantom limb effects are just as real in sensation space although they have no correspondence to outside physical events. Pain is, I believe, not caused by physical events but is associated with them. Pain is purely a sensation-space phenomenon resulting from operational transformations of intuitions into sensations. By transforming intuition, sensation can be created and transformed.

Now what is a well-defined intuition? The *American Heritage Dictionary of the English Language* defines *intuition* as *the act or faculty of knowing without the use of rational process; immediate cognition; a sense of something not evident or deducible. Sharp insight.* Its Indo-European root is *teu,* which means to pay attention to, to look at.

An intuition is not a wishy-washy sentiment or feeling. It has definite consequences. One could think of an intuition as a program of particular sensations, particularly sensations that could occur in the future. Since we operate in sensation space, we interpret all of our intuitions in terms of sensations. The symbolic organization of sensations results in thoughts or the temporality of mental experience.

The more defined a sensation, the greater is the unpredictability of future sensations. Sensations transform into intuitions and back again as a result of our thinking and observing. The great discovery to come will be the "lens" that accomplishes this transformation. It may be nothing more than our language.

If the words we use are our "lenses," they shape and transform our sensations, altering our internal quantum wave functions and providing us with intuitional senses of the future. Intuition cannot be rationalized because the principle of complementarity acts within our nervous systems.

In the next section we will look at the mathematics of the transformation of thoughts into feelings and the inverse transformation back again.

The Mathematics of the Transformation of Feelings into Thoughts and Back Again

Just as energy and time play the game of uncertainty (see page 270), there is a similar complementarity of feelings and thoughts. However, just as energy is observable in terms of its changes, which mark and give rise to time, feelings are operationally observable, and their changes are the markings of thought. Just as time turns out not to be operationally observable, it also turns out that thoughts are not operationally observable. If you try this moment to watch your thoughts, you will quickly realize that there is no way to do this without recording them. To grasp this difficult concept we shall again need mathematics and the mathematical functions that transform energy into time and back again. These functions in quantum mechanics are called *time-displacement operators*. In a similar manner we will be looking at *thought-displacement operators*.

As I have pointed out in Chapter 1, time cannot be operationally observed. We cannot decide to look for an event at any time. Once time has passed, it has passed. However, space does not hold that barrier for us. We can look both forward and backward in space but not so in time. In our new language, we say that time cannot be represented by a ket, or a state function. In other words, time is not an observable in quantum physics. Instead it is inferred from observations describing a sequence of position measurements of an object. In a typical sequence recorded in a film emulsion or bubble chamber, a particle leaves a track of spots or bubbles. By measuring the lengths of these tracks and knowing the momentum of the particle (by preparing the beam of particles fired toward the emulsion), one infers that the particle was at a certain place at a certain time. All one actually measures is the distance between the spots or bubbles.

In an ordinary experiment an electron sequentially collides with atoms of a film emulsion or bubble chamber. Since the electron has a well-defined momentum, it has no definitive location in space. When it hits one atom, this momentum quantum wave function is scattered, changing its shape but not its momentum (the wave changes from a plane wave into a wave that spherically expands away from the scattering atom acting as the center of the newly formed spherical wave). When this spherical qwiff reaches another atom, a second scattering takes place. After many such scatterings a track is left behind because the centers of these spheres left marks on the film or in the bubble chamber.

These changes are nothing more than the transformations from momentum to position, to momentum, to position, and so on, mentioned in the previous section of this chapter. The track of the transformations infers the passing of the electron from one atom to the next *in time*.

Even though this passing cannot directly describe the time of an event, quantum physics uses a parameter to describe the movement of a qwiff or a ket from one state to another. This parameter is called the time-independent variable and this change in a qwiff or a ket is called its *time development.*

Closely associated with the time development of a ket is the energy of the system described by the ket. From the uncertainty principle we know that it is not possible to describe a state with a well-defined energy and have that state begin and end within a specified time interval. Similarly, any system that has a specified time interval (i.e., a beginning and an ending time) will not have a well-defined energy. In quantum physics energy is associated with the wave frequency, or number of vibrations per second, of the qwiff representing the system or particle under question. This relationship is written (see Chapter 8, page 210)

$$E = hf$$

where E is the energy of the particle, h is Planck's constant, and f is the frequency of the wave. Since E is an observable of quantum physics, it has an associated operator called the Hamiltonian (after Lord Hamilton, see Chapter 3, page 102) or H operator. This H-op also follows Law III of Chapter 8, namely, that there exists a set of kets each of which is associated with a different energy:

$$ei, \ i = 1, 2, 3, \ldots$$

These kets are labeled as ket-Ei's, and for them Law III becomes

$$<I|H|Ei> \ = \ ei <I|Ei>$$

so that when H-op is sandwiched between any I-bra state and an energy state ket, ket-Ei, it produces a value, ei, for the energy associated with the ket-Ei. The quamp $<I|Ei>$, remember, is the question: If the state of the system is energy state Ei, then is it in state I?

The existence of energy state Ei means that the system has a certain persistence of experience, that is, a definite energy. This means that there is a stable oscillation or constant frequency associated with that experience.

290

Here the time development of a ket enters. If there is a constant energy or frequency, then the ket is said to be in a stationary state of oscillation. Nothing is changing in time because there is no time to mark the beginning and the end of the oscillation. One says that the state ket is "rotating" in the abstract Hilbert space, which is the ideal space of all energy possibilities.

To describe this rotation a time-development or time-displacement operation is needed. The operator for this is more abstract than any of the previous operators so far encountered. It is called the time-displacement operator and is responsible for the creation of time experience. This operator acts as a scanner, stretching along the wave form in either direction.

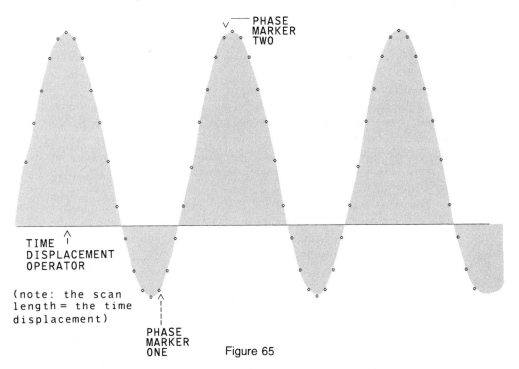

PHASE
MARKER
TWO

TIME
DISPLACEMENT
OPERATOR

(note: the scan
length = the time
displacement)

PHASE
MARKER
ONE

Figure 65

The question for the scanner is, What is the frequency of the wave? In order to determine this, the *scan-length* cannot be too short. It must be long enough to sample at least one complete vibration, one period. The longer the scan, the greater the number of periods sampled. But therein lies the rub. The increased length of the scan-length provides many repeats of the vibration. They appear as a single note held very, very long, as if sung by a powerful contralto. In this long note there isn't a clue as to when the note began or when it will end. A single frequency has no clearly marked beginning or ending.

However, if there is a change in frequency, that change appears as a time marker. Just as when the contralto began to sing the concert started and marked $t = t_0$, a change in notes marks the end of a cadenza, $t = t_1$. Any phase difference is the potential marking of a real time interval.

Similarly if there are several changes in frequency, there will "appear" several marking points, $t = t_2, t_3, \ldots$, each marking the beginning and the ending of a time interval. Thus it is that time appears to us as a change in frequency, which is a change in energy of the state of a system.

It is possible to experience an analogy of the time-displacement operator by listening carefully to a trill played on the keyboard of a piano. When, for example, the two lowest adjacent notes are played too rapidly, it is impossible to hear the change in pitch. If the same two notes are played just as rapidly several octaves higher, we have no difficulty hearing their difference in pitch.

The time operations performed by the rapidly moving fingers cannot clearly differentiate one low note from the other because the notes themselves have not been played long enough for the ear to mark distinctly their separate frequencies. This presents no problem for the high notes because enough vibrations per second have reached the ear before the notes are changed. The time interval as given by the change in frequency must be longer than the time it takes for one note, one wave period passing, to be distinctly identified. Thus there has been no identifiable change in notes in the low-frequency case even though the trill was played. In this sense no time displacement has occurred for the low-note trill, while a clear time interval has been marked by the high-frequency trill.

Thus it is that a state with a well-defined frequency, one that has a well-defined energy, cannot be a measure of a beginning and an ending time. A time-displacement operation performed on it will not yield a new event. Although there is repetition, there is no time displacement or time interval. However, if two or more energy states are added together and the time-displacement operation is carried out on the sum, there will appear relative differences in phase. These relative phase differences can serve as time markers.

In fact, the time-displacement operator is itself made up of a sum of phase advances for different energy states, that is, different frequencies. To see how the time-displacement operation occurs we need to look at energy states and energy state kets.

An abstract constant energy state can be represented in space as a quamp:

$$\langle x | E \rangle$$

that poses the question: If the state has energy E, then is it located at x? This quamp transforms the ket-E to a quantum wave function in physical space. Thus $\langle x | E \rangle$ is a mathematical function of x, like $y(x)$, that exhibits a characteristic shape dependent on the value of E. For a free particle this function is the same as a momentum state, $\langle x | p \rangle$, and in fact there is a simple relationship between the momentum, p, and the energy, E, of a free particle:

$$E^2 = p^2c^2 + m^2c^4$$

where m is the mass of the free particle when it is at rest and c is the speed of light.

As before, the inverse of $<x|E>$:

$$<E|x>$$

asks the inverse or conjugated question: If it happened at x, then does it have energy E? This quamp transforms the location of an event in space to an occurrence with an energy E.

The state $<x|E>$ has a well-defined energy, so it is not marked in time. However, just as it is possible to represent a momentum state as a sum of position states, and a position state (the Dirac delta function) as a sum of momentum states, it is possible to represent the transformation in time of a position state to another position state by using different energies, e_i, and correspondingly different energy states, $<x|E_i>$, and their inverse transformations, $<E_i|x>$. By adding together, "trilling," a number of energy states, any arbitrary state not having a well-defined energy can be represented. By giving up information about energy, markings of beginnings and endings of time intervals can be obtained.

To do this one must first advance the phase of each wave over the desired time displacement for each energy state:

$$PA(t2, t1; E_i)$$

PA is a mathematical function that marks the phase advance from time $t1$ to time $t2$ along the well-defined energy state E_i. This advance occurs for each energy state, but not yet as an observable time displacement.

The time-displacement operator is then computed as a sum of different energy states in the following manner.

Suppose we know that the location of a particle at time $t1$ is $x1$. Its quamp is a Dirac delta function (see Figure 64), only this time its ket is marked in time at $t1$:

$$<x1|x1, t1>$$

equal to zero along the x direction of space everywhere except at $x1$. To find out where the particle will be later we must first transform the quamp from x space to energy state E_i, as

$$<E_i|x1> <x1|x1, t1>$$

(remember to read this formula from right to left as in Hebrew) and then let the phase advance function, $PA(t2, t1; E_i)$, advance the phase of the energy state Ei from the phase associated with beginning time $t1$ to the phase associated with time $t2$:

$$PA(t2, t1; Ei)<Ei\,|x1> <x1\,|x1, t1>$$

Then we must transform back to the energy state Ei, but at the new position, $x2$, in space:

$$<x2|Ei> PA(t2, t1; Ei) <Ei\,|x1> <x1\,|x1, t1>$$

and then transform from energy state Ei back to space and end up with the particle at location $x2$ at time $t2$:

$$<x2, t2|x2> <x2|Ei> PA(t2, t1; Ei) <Ei\,|x1> <x1\,|x1, t1>$$

Finally we must "trill," or add together, all possible energy states Ei that mark the possible energy with which the particle located at $x1$ at time $t1$ could reach the position $x2$ at time $t2$:

$$<x2, t2\,|x1, t1> = \text{SUM (over all energy states } Ei) \text{ of}$$
$$<x2, t2|x2> <x2|Ei> PA(t2, t1; Ei) <Ei\,|x1> <x1\,|x1, t1>$$

The result of all this is the time-displacement operation

$$T(x2, t2; x1, t1) = <x2, t2\,|x1, t1>$$

or the quantum probability amplitude that a particle located in space-time at $x1$, $t1$, "made it" to space-time location $x2$, $t2$.

Figures 66A–F display the time-displacement quamp for various times, $t2$, for a particle initially located in the center of the paper at starting time $t1 = 0$.

In Figure 66A we see a broad hill with a gentle falling off to valleys with falls and rises. The frequency of these falls and rises increases progressively as you go farther out on either side. This curve corresponds to the particle after the longest time has passed from the start. Figure 66B corresponds to an earlier time. It has a slightly higher and narrower central hill and begins to form higher-frequency oscillations closer in toward the central hill. As your eye progresses through Figures 66C, D, E, and F, corresponding to earlier and earlier times, the peaks become higher, the central hill becomes higher and narrower, and the rapid oscillations begin closer to the central hill. In all of the curves, the farther one is from the central hill, the more rapid is the variation in terrain.

At the earliest time, Figure 66F shows a smooth curve over a short range of space that begins to oscillate rapidly as one moves only a short distance away from the center. These rapid oscillations correspond to higher and higher frequencies. The farther from the center one looks for the particle, the greater the oscillation frequency. (I wasn't able to draw these rapid oscillations more accurately because too many had to be drawn within the width of the line made

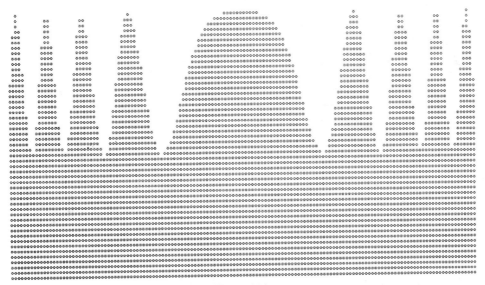

Figure 66A

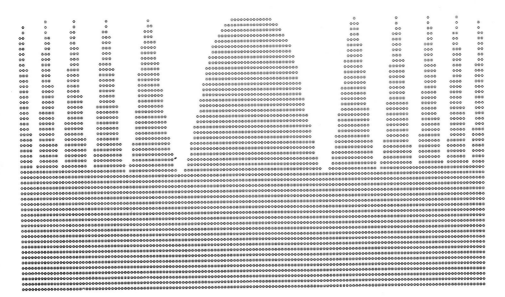

Figure 66B

Figure 66C

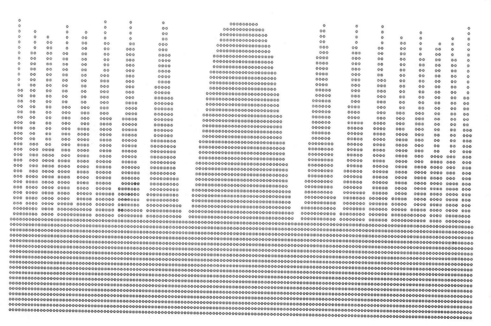

Figure 66D

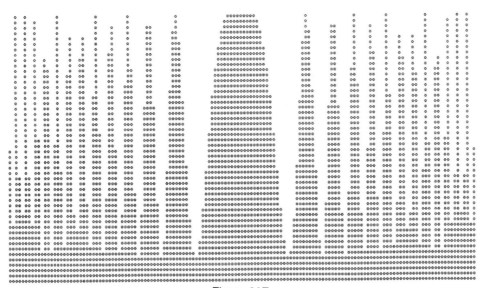

Figure 66E

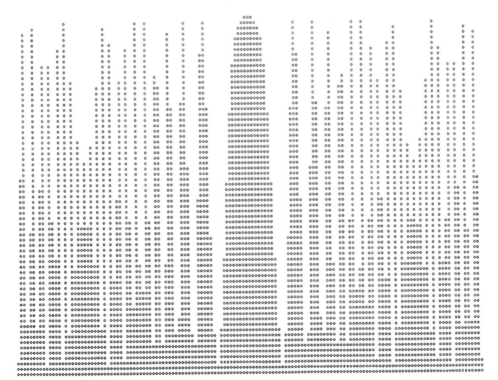

Figure 66F

297

by my computer.) This increase of frequency with greater distance means a corresponding increase in energy, which is necessary in order to get the particle "out there" in the short time, $t2 - t1$. The farther away from the center, the higher the energy needed to get it there in the time interval $t2 - t1$.

If any of the drawings were extended in space to the left and right, the frequency of the oscillation would increase. More and more oscillations would appear, with the peaks becoming more and more crowded together. If one looked at a region of space well to the right, for example, the extreme crowding of the oscillations would produce such a rapid oscillation that the probability within that region would average out to zero. This is similar to attempting to swing an object back and forth very rapidly. The object would be pushed and pulled so many times that its responses would be not to move at all.

In this manner we can see that the space around the particle (by which I mean the possible location of the particle outside the immediate hill range) is filled with extremely high-frequency oscillations too rapid to be sensed. The result is that no particle is found too far away from the center.

As time goes by, as $t2$ is increased, the flat part of the quamp spreads out farther, "pushing" the high-frequency vibrations farther out. This spreading of the quamp means that with time there is a loss of information associated with finding the particle within any given spatial interval. Here we have the idea that time measures the information loss in space, or the increase in entropy.

In a similar manner we can consider a complementarity between feelings and thoughts. Feelings are observable in terms of their changes, which mark and give rise to thought. Feelings are therefore operationally observable, and their changes are the markings of thought; while thoughts are not operationally observable. You cannot observe them without disturbing them. To describe the transformation of feeling into thought and back again we can use the same mathematical functions that transform energy into time and back again. These functions are the thought-displacement operators.

As I have pointed out, thought cannot be operationally observed. Once thought has passed, it has passed. Sensation does not hold that barrier for us. Just as we can move to the left or right in space, we can observe sensations in sequences that, because of their simultaneity, are not in a unique causal order (see the discussion of spacelike separated events, Chapter 7). Thoughts require time and causality. Thoughts cannot be spacelike. The acausality of our sensations causes the difficulty we all have in telling the doctor about our ills, in explaining our "reasons" for acting as we did in crises, and, in general, in describing our "feelings."

In our new language we say that thought cannot be represented by a ket, or a state function. In other words, thought is not an observable in quantum psychodynamics. Instead it is inferred from observations describing a sequence of sensations within the nervous system. In a typical sequence recorded in our nervous systems, an event leaves a track, a set of engrams. By observing the sequences of these tracks and knowing the intuition state of the thought (by remembering or having a similar set of sensations), one infers that the experience was a certain event corresponding to a certain thought. All one actually observes, however, is the sensation sequence produced by the "refired" set of engrams.

In an ordinary thought experience, potassium or possibly calcium ions sequentially interact with atoms of a neuronal gate. Since these ions have well-defined momenta, they have no well-defined location in space. When one of them hits one atom, its momentum quantum wave function is scattered, changing its shape but not its momentum (the wave changes from a plane wave into a wave that spherically expands away from the scattering atom acting as the center of the spherical wave). When this spherical qwiff reaches another atom, a second scattering takes place. After many such scatterings a track is left behind because the centers of these spheres made marks in the synapses.

These changes are also nothing more than the transformations from momentum to position, to momentum, to position, and so on, mentioned in the previous section of this chapter. The track of the transformations implies the passing of the ion from one atom to the next and is the root of thought.

Even though this passing cannot directly describe the thought of an event, it is possible to describe the timelike sequence of events that are thoughts through the use of a similar "thought" parameter describing the movement of a qwiff or a ket from one state to another. This parameter, like the time-independent variable that was mentioned earlier, can be called the thought-independent variable, and this change in a qwiff or a ket can be called its *thought development.*

Closely associated with the thought development of a ket are the feelings of the system described by the ket. By extending the uncertainty principle from energy-time into feelings-thought, I theorize that it is not possible to describe a state of mind having well-defined feelings *and* having that mental state begin and end with a specific thought. Similarly, any mind that has a specific thought will not have well-defined feelings. Just as in quantum physics energy corresponds to a frequency, in the "new psychology" we can associate with feelings a wave frequency or number of vibrations per second of the qwiff representing the psychic event. This relation is written:

$$F = hf$$

where F is the feelings of the event, h is Planck's constant, and f is the frequency of the wave (see Chapter 8, page 210). Since F is taken to be an observable of quantum physics, it will also have an associated operator that I call the Jungian (after Carl Gustav Jung), or the J operator. This J-*op* also follows Law III of Chapter 8, namely, that there exists a set of kets each of which is associated with a different feeling:

$$fi, \quad i = 1, 2, 3, \ldots$$

These kets are labeled as ket-*Fi*'s, and for them Law III becomes

$$<I|J|Ei> = fi <I|Fi>$$

so that when the Jungian operator, *J-op*, is sandwiched between any *I*-bra state and a feeling-state ket, ket-*Fi*, it produces a value, *fi*, for the feeling associated with the ket-*Fi*. The quamp $<I|F>$ is the question: If the nervous system is in feeling state *Fi*, then is it in state *I?*

The existence of feeling state *Fi* means that the person has a definite feeling or a certain persistence of experience. This means that there is a stable oscillation or constant frequency associated with that experience.

Here the thought development of a ket enters. If there is a constant feeling or frequency, then the ket is said to be in a stationary state of oscillation. No thought occurs, because there is no thought to mark the beginning and the end of the oscillation. One says that the state ket is "rotating" in the abstract Hilbert space, which is the ideal space of all feeling possibilities.

To describe this rotation the thought-development or thought-displacement operation is needed. The operator for this parallels the previous time-displacement operator in form. It is called the thought-displacement operator and can be viewed as being responsible for the creation of thought experience. This operator acts as a scanner, moving along the wave form in either direction similar to the time-displacement operator shown in Figure 65.

As the scanner moves along a quantum wave with constant frequency it continues to "see" the same wave again and again but at different places or phases along the wave. In this manner it cannot determine that any thought occurs because the repetitions of the wave do not mark any change. The frequency is unchanging, so thought is not defined. No reference point exists. If, on the other hand, there is a change in frequency, the location of the change serves as a marker for the beginning of thought. Similarly, if there are several changes in frequency, there will "appear" several marking points, each marking the beginning and the ending of a thought. Thus it is that morphemes appear to us as changes in frequency, which are changes in the feelings of the person.

Thus it is that a person who has a well-defined feeling, a well-defined frequency, cannot experience thought. A thought-displacement operation will not yield a new thought. Although there is repetition, there is no thought displacement — no thinking is taking place. However, if two or more feeling states are added together and the thought-displacement operation is carried out on the sum, there will appear relative differences in phase. These relative phase differences serve as thoughts.

The thought-displacement operator is itself made up of a sum of phase advances for different feeling states, that is, different frequencies. To see how the thought-displacement operation occurs we need to look at feeling states and feeling-state kets.

An abstract constant feeling state like an abstract constant energy state can be represented in terms of sensations — in a sensation space as a quamp:

$$<s|F>$$

which asks the question: Does feeling F produce sensation s? This quamp transforms the ket-F to a quantum wave function in sensation space. Thus $<s|F>$ is a mathematical function of s, like $<x|E>$, that exhibits a characteristic shape dependent on the value of F.

Just what the transformation is is difficult to say. By analogy there must exist some "parallel" within the nervous system between a morpheme and a particle. Just as a particle has a rest mass that is trapped light energy, a morpheme consists of trapped feelings. The words we use are composed of morpheme sequences in the same sense that our neural bodies are composed of encoded sequences of molecular strings. A morpheme, m, in the nervous system may act in relationship to intuition and feeling as the rest mass, m, of a free particle acts in relationship to momentum, p, and energy, E. If so, the functional relationship between feeling, F, intuition, q, and morpheme, m, might satisfy a similar mathematical equation, namely

$$F^2 = q^2c^2 + m^2c^4$$

and in fact this is the same relationship between the momentum, p, and the energy, E, of a free particle, where m is the mass of the free particle when it is at rest, and c is the speed of light.[†]

To what the invariant speed of light, c, corresponds in the nervous system is speculative. I suggest that it has something to do with the "speed of thought" or the rate at which morphemes can be assembled into words. Just as one particle contains an enormous trapped energy, one morpheme contains enormous trapped feelings. Ordinary languages tend to inhibit the liberation of those feelings just as ordinary matter tends not to release its trapped energy ($E = mc^2$). There is a stability to morphemes. This stability may manifest as fundamental "sounds" in the same sense as matter manifests as fundamental particles. This speculation lends credence to the ancient laws of the Qabala and the biblical pronouncements of John: "In the Beginning Was the Word."

As humans have learned to tap the mc^2 of particles and liberate their trapped energies, sacred languages, hymns, and other mystical incantations

[†]I realize that even by writing this I may be accused of a neomechanical reductionism. Again, I point out that quantum physics already rids me of this taint. A physics of the psyche is no more reductionistic than the quantum physics of the universe. If anything, my work emphasizes an almost neomysticism. It attempts to explain the difficulties we all have in understanding our feelings. Just how sensations transform into feelings and back again becomes, in this attempt, no more (or less) mysterious than the transformation of energy into the physical location of particles. Both are quantum processes.

may also liberate the feelings trapped in morphemes.[†] In fact the rise and fall of civilizations may be as dependent on these ritualistic or religious experiences as the modern technosociety is dependent on nuclear weaponry. By regaining our sacred heritage of the use of morphemes we may be able to counter the forces of destruction threatening us. These forces, for good and evil, for creation and destruction, arise as the physical and psychic manifestations of the uncertainty principle. Feeling/thought is the other side of the coin of energy/time.

For a free morpheme the feeling-wave-in-sensation-space quamp $<s|F>$ is the same as an intuition state, $<s|q>$.

As before, the inverse of $<s|F>$:

$$<F|s>$$

asks the inverse or conjugated question: Does sensation s produce feeling F? This quamp relates the sensation of an event to a state with a feeling F.

The state $<s|F>$ has a well-defined feeling, so it is not marked in thought. However, just as it is possible to represent an intuition state as a sum of sensation states and a sensation state as a sum of intuition states, it is possible to represent the transformation by thought of a sensation state to another sensation state by using different feelings, f_i, and correspondingly different feeling states, $<s|F_i>$, and their inverse transformations, $<F_i|s>$. By adding together, "trilling," a number of feeling states, any arbitrary state that does not have a well-defined feeling can be represented. By giving up information about feelings, markings of beginnings and endings of thought intervals can be obtained.

To do this one follows a procedure similar in form to computing the phase advance from time $t1$ to time $t2$ along a well-defined energy state. Simply replace energy E_i with feeling F_i and times $t1$ and $t2$ with thought markers $t1$ and $t2$. This produces a thought-displacement operator, TPA, for each feeling state F_i:

$$TPA(t2, t1; F_i)$$

The thought-displacement operator is then computed as a sum of different feeling states in the same manner.

Consider a morpheme associated with a sensation, $s1$. This morpheme marks the start of a thought. This thought may be only a simple word or syllable. I write this event as occurring as thought $t1$. I use the word *thought* as a general concept, not as a specific thought one might have. Thought is to morpheme occurrences with sensations as time is to particle occurrences as events in physical space.

[†]Indeed, many chants and mantras consist of "meaningless" sounds repeated over and over again.

The quamp for a simple morpheme produced as a starting thought, $t1$, at sensation $s1$ in sensation space appears to the nervous system as a sensation, $s1$. Its quamp is written:

$$<s1|s1, t1>$$

This sensation is the primal nervous experience of the start of a word. It is the abstract primary sensation event, the primitive morpheme. By stringing together morpheme sequences, we form actual words. The thoughts we utter or think to ourselves are created "all at once," before they are spoken or sensed as words. The following describes the quantum-physical, that is, quantum-psychic, processes for this formation to occur.

By starting thought $t1$, I mean the abstract process associated with any sensation that marks the morpheme. A morpheme sensation is propagated into thoughts — reappears as other sensations — by first transforming the quamp from s space to feeling state Fi as:

$$<Fi|s1> <s1|s1, t1>$$

Second, the phase of the feeling state Fi advances from the phase associated with beginning thought $t1$ to the phase associated with thought $t2$:

$$TPA(t2, t1; Fi) <Fi|s1> <s1|s1, t1>$$

Third, it is transformed back to the feeling state Fi, but at the new sensation, $s2$:

$$<s2|Fi> TPA(t2, t1; Fi) <Fi|s1> <s1|s1, t1>$$

Fourth, this is transformed from feeling state Fi back to the morpheme with sensation $s2$ at thought $t2$:

$$<s2, t2|s2> <s2|Fi> TPA(t2, t1; Fi) <Fi|s1> <s1|s1, t1>$$

Finally we must "trill" or add together all possible feeling states Fi that mark the possible feeling with which the morpheme located at $s1$ at thought $t1$ could reach the sensation $s2$ at thought $t2$:

$$<s2, t2| s1, t1> = \text{SUM (over all feeling states } Fi) \text{ of}$$
$$<s2, t2|s2> <s2|Fi> TPA(t2, t1; Fi) <Fi|s1> <s1|s1, t1>$$

The result of all this is the thought-displacement operation

$$T(s2, t2; s1, t1) = <s2, t2| s1, t1>$$

or the quantum probability amplitude that a sensation, $s1$, associated with the start of thought $t1$, ended with a sensation, $s2$, and the end of the thought, $t2$.

303

Figures 66A – F also display this quamp for the propagation of a single morpheme through a range of sensations at different thoughts, $t2$. In the nervous system the free morpheme reaches into sensation space. The more sensations it encompasses, the less likely is the morpheme associated with any particular sensation. As thought progresses, the probability for the morpheme to be associated with a particular sensation decreases. The morpheme, described as a primary event in sensation space, is initially located in the center of the paper at the start of thought $t1 = 0$ in Figure 66F.

The result is that with increased thought or time the morpheme is spread over a wider range of sensations with a corresponding loss of location of the initial sensation marking the morpheme. This loss is experienced as the beginnings of speech and thought, and with nothing further happening no word or thought is actually formed. This is nothing more than an animal yelp or grunt.

As thought goes on, as $t2$ is increased, the flat part of the quamp spreads out farther, "pushing" the high-frequency vibrations farther out. This spreading out of the quamp means that with greater thought there is a loss of information associated with finding the morpheme within any given range of sensations. It is here that I offer the idea that thought is a measure of information loss in sensations, that is, an entropy increase occurs. The inability to associate the initial morpheme with any range of sensations increases with thought. Thus thoughts appear to have no defined associated body sensations.

The morpheme thought-displacement operator need not act only on a single morpheme. By starting with a quantum wave of sensations, one uses the thought-displacement operator to turn the wave into thought. Our thoughts, which "just occur to us," are later turned into spoken or written words as the physical manifestations of the thought-displacement operation performed on a vast and complex quantum wave form of many morphemes.[†]

The spread of a morpheme over a range of sensations is needed to provide human speech. The ability to speak is the ability to think. Speech consists of physically manifested thoughts utilizing the vocal mechanism. Writing also consists of manifested thoughts using the mechanism of the hand and fingers. You could think of the vocal chords and the fingers as the physical manifestations of thoughts, where thought is the spread of morphemes into physical sensations.

The rapid alteration of operations created by observations of sensations produces words. In the parallel worlds theory any hill in Figure 66 is a starting hill for a parallel worlds jaunt. From that hill a new morpheme spreads out in its parallel sensation spaces. The overlap of these worlds either produces a resonance or it doesn't. This resonance is the formation of a single word. When you speak, the whole human race from the beginning of time echoes in your voice. (See Figure 67.)

[†]This is quite similar to the Feynman path integral technique used in modern physics to describe the many-body problem.

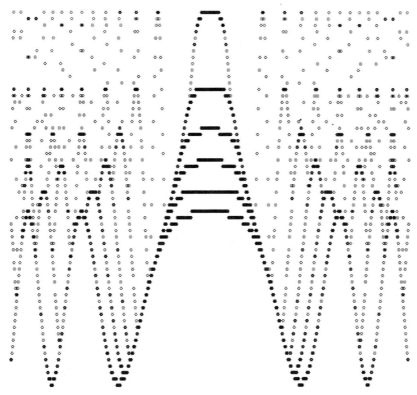

Figure 67

The echoes of time. The above shows a single morpheme propagating over a range of sensation space. The latest time is shown by the front curve and the earliest by the rear curve. This spreading out of a single morpheme is predicted by quantum physics in the same manner as a particle spreads out in physical space after an initial discovery. A single sound generates a range of sensations. It is the coordination of these sensations that results in human speech.

The thought-displacement operator transforms feelings into thought by spreading a morpheme over a wider range of sensations. This is captured in the parallel physical plane as human speech. Our bodies are embodied feelings. They are the physical mechanisms for producing these transformations. The expression of feeling is the physical production of sound, facial expression, body posture, etc. Thus our bodies embody trapped feelings. By thinking, we transform feelings into sensations in the same way that a particle propagates from one place to another.

The transformation operation, $T(s2, t2; s1, t1)$ depends on the relationship between feeling, F, intuition, q, and morpheme, m. For a free morpheme this relationship is simple. However, just as there are physical situations that act as inhibitors of free movement on the physical plane, there are correspond-

305

ing inhibitions that limit the spread of a morpheme over sensations. Patterns of thinking, like the proverbial needle stuck in a record groove, tend to be reinforced by other forms of relationship between feelings, intuition, and morpheme.

Repeated Thoughts and Feelings

For a physical system that is bound to oscillate harmonically, the relationship between energy, momentum, and mass also involves the physical location of the particle. It is written

$$E = p^2/2m + mw^2x^2/2$$

where E is the particle's energy, m is its mass, p its momentum, x its location in space, and w the frequency with which it vibrates back and forth if it is in a given energy state.

In Figures 68A – U we see how a harmonically bound particle propagates. Actually, just preceding 68A there is a condition consisting of a delta function in the middle of the page with infinite frequency oscillations on either side. Figure 68A shows the transformation at a slightly later time. As time progresses, the quamp shows a similar progression as is shown in Figure 66; but instead of a continual spreading, we see that in Figure 68F the quamp is spread with equal probability over the whole range of space in a finite period of time. This means that the particle is "anywhere." As time goes on, the reverse occurs as the quamp propagates back into itself. Between Figures 68K and 68L the delta condition is reached again. The particle is suddenly collapsed and appears in the center of the figure. After this the pattern repeats with a frequency characteristic of the energy E.

This is the continual transformation of E from position energy (energy associated with x, having a location in space) into momentum energy (or wave energy with no location in space) and back again. The total overall energy remains constant but not during a single cycle of transformation.

For a psychic system that is bound to oscillate harmonically, the relationship between feeling, intuition, and morpheme also involves the sensation location in the body. It is written

$$F = q^2/2m + mw^2x^2/2$$

where F is the feeling energy, m is the morpheme "mass," q is the intuition quality of the morpheme, x is its sensation location in the body, and w is the frequency with which the sensation vibrates back and forth if it is in a given feeling state.

Looking again at Figures 68A – U, we see how a harmonically bound morpheme propagates starting with Figure 68A. Instead of continually spreading we see that in Figure 68F the quamp is spread with equal probability over the whole range of sensation in a finite thought period. This means that the morpheme is "anywhere." The sound fills the whole body and a sense of knowing occurs. As thought goes on, the reverse occurs as the quamp propagates back into itself. The morpheme is suddenly collapsed and appears in the center of the figure, where a single sensation is definitely felt or repeated. After this the pattern repeats with a frequency characteristic of the feeling *F*.

This is the continual transformation of *F* from sensation feelings (feelings associated with *s*, having a defined sensation) into intuition feeling (or wave feeling with no body sensation) and back again. The total overall feeling remains constant but not during a single cycle of transformation.

These transformations involving bound feelings tend to oscillate according to the uncertainty principle. Transformation from feelings into intuition or its conjugate sensation repeats itself with thought, which appears through time. The overall feeling is the same, but its manifestation in the body depends on the binding force tending to hold the morpheme "in place." Our bodies reflect these tendencies.

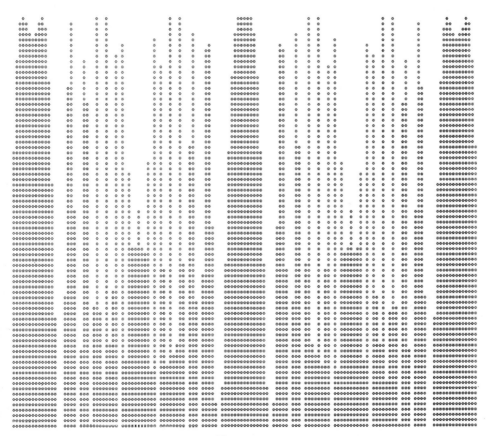

Figure 68A

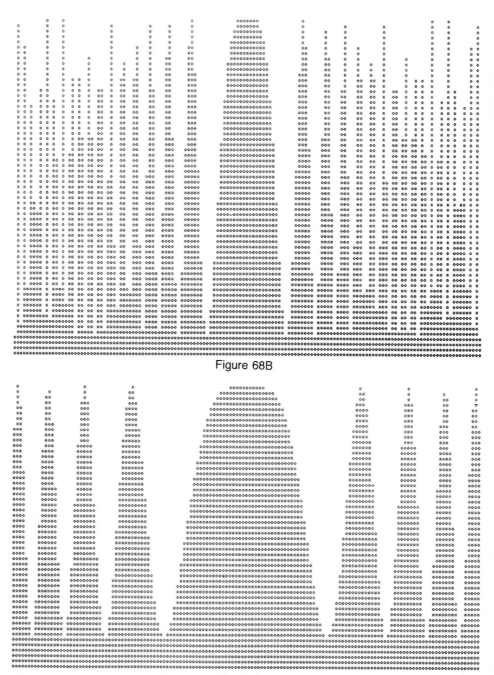

Figure 68B

Figure 68C

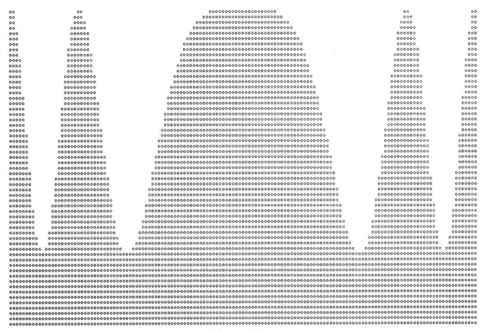

Figure 68D

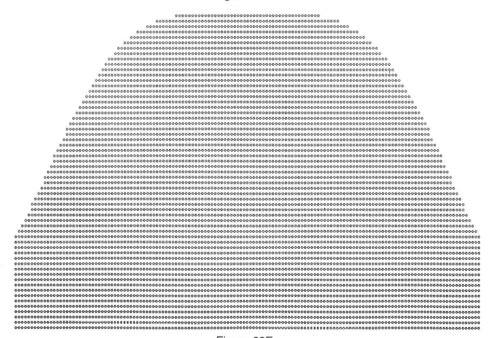

Figure 68E

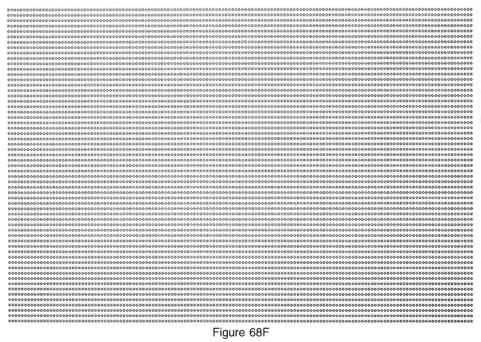

Figure 68F

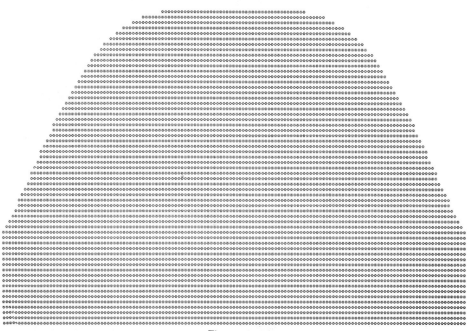

Figure 68G

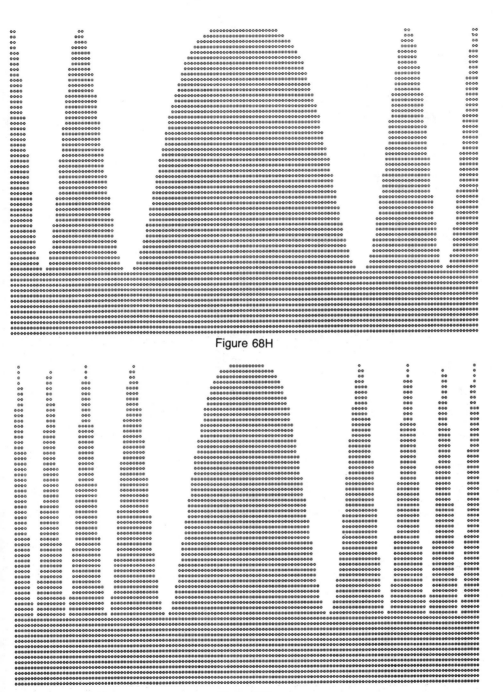

Figure 68H

Figure 68I

311

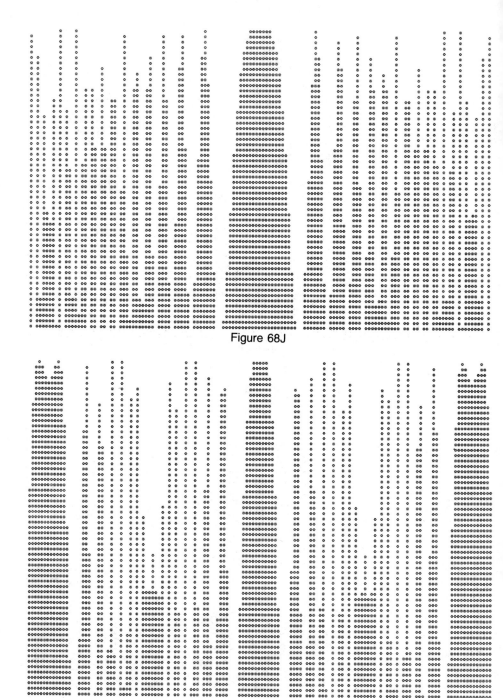

Figure 68J

Figure 68K

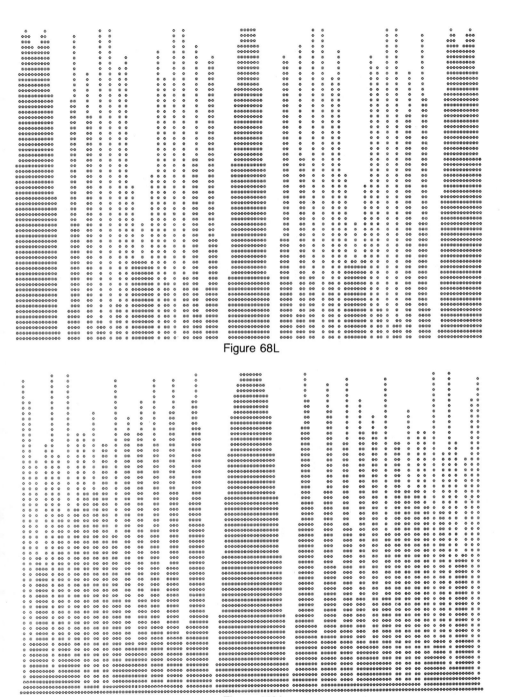

Figure 68L

Figure 68M

Figure 68N

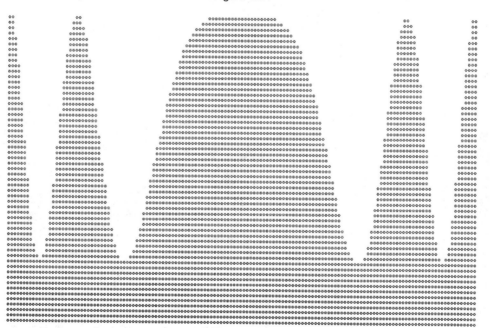

Figure 68O

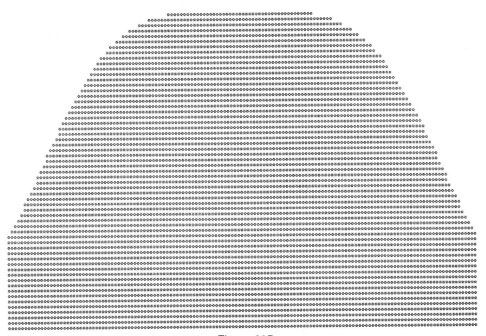

Figure 68P

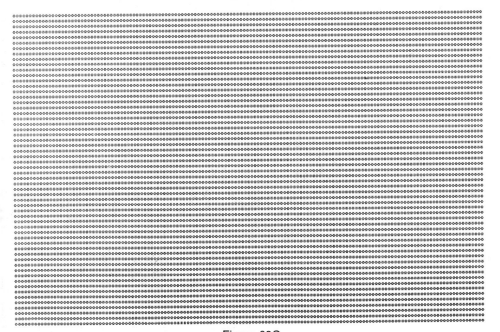

Figure 68Q

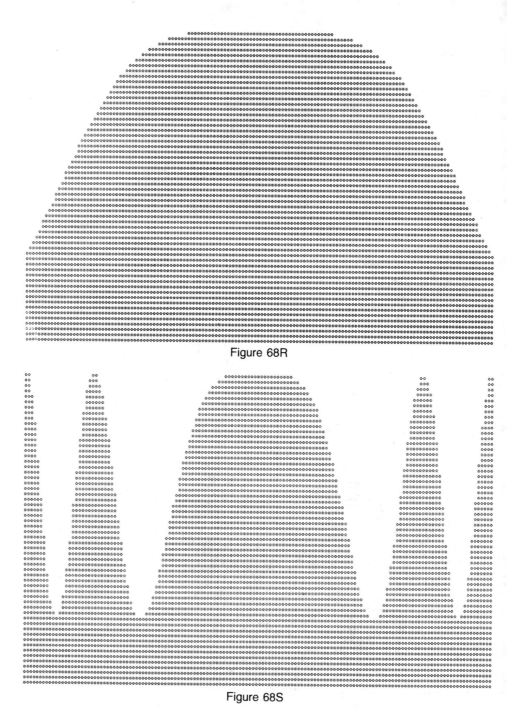

Figure 68R

Figure 68S

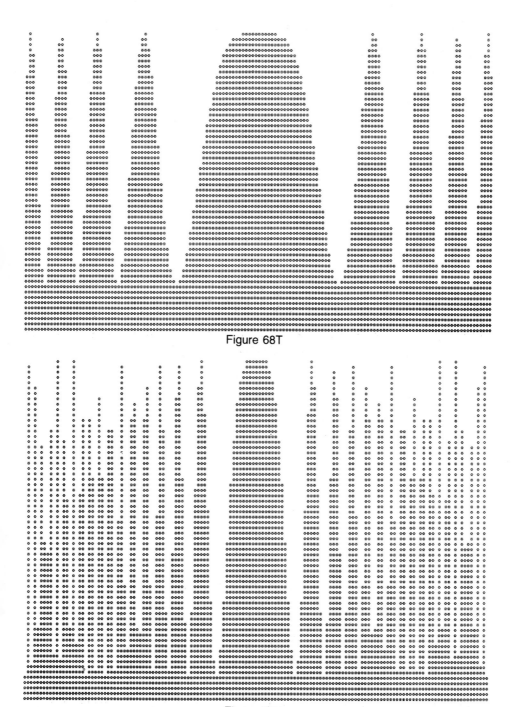

Figure 68T

Figure 68U

Memory in the Quantum Holographic Parallel Worlds Model

From the above we have seen what it means to have alongside our everyday observations of physical events a parallel sequence of feelings, thoughts, intuitions, and sensations. To complete the picture we need to explore how these two worlds interact. The holographic model of the brain together with the parallel worlds model of quantum physics provides a highly plausible answer.

Now that we have looked at the idea of a holographic mind (see Chapters 6 and 9) and have explored the idea of sensation space as a parallel to physical space, it is necessary to enquire into the quantum mechanism of memory. Just why do we feel the ways we do when we see or sense certain external experiences? I have pointed out earlier that our brains record information holographically. I have used the notation $[IW]$ to represent the complex wave of information that reaches our brains and *somehow* becomes a record or engram.

> By interfering with a reference wave $[RW]$ and taking into account that the energy deposited in a holographic medium is the product of the wave pattern, $[W]$, with its complex conjugate, $[W]^*$:
>
> $$\text{Energy deposited} = [W] \times [W]^*$$
>
> we found that the sum of the reference wave $[RW]$ and the information wave $[IW]$:
>
> $$[RW] + [IW]$$
>
> produced a recording of
>
> $$([RW] + [IW]) \times ([RW]^* + [IW]^*)$$
>
> which resulted in four terms:
>
> $$[RW]^*[RW] + [RW]^*[IW] + [IW]^*[RW] + [IW]^*[IW]$$
> $$\text{term 1} \qquad \text{term 2} \qquad \text{term 3} \qquad \text{term 4}$$
>
> In terms 1 and 4, because they were each "mirror image" multiplications, all phase information was lost. The recording of these terms produces a uniform "clouding" of the medium but no replayable information. Terms 2 and 3 contain the desired information, which is replayed when a new wave is transmitted through the medium. In this way the medium acts as a filter.
>
> When the new wave is $[RW]$, it interacts with the medium by simply multiplying its wave amplitude by each of the terms present:
>
> $$[RW][RW]^*[RW] + [RW][RW]^*[IW] + [RW][IW]^*[RW] +$$
> $$\text{term 1} \qquad\qquad \text{term 2} \qquad\qquad \text{term 3}$$
> $$[RW][IW]^*[IW]$$
> $$\text{term 4}$$

This yields a replay of term 2, with the other terms "clouding" out. If the new wave is $[RW]^*$, giving:

$$[RW]^*[RW]^*[RW] + [RW]^*[RW]^*[IW[+ [RW]^*[IW]^*[RW +$$

| term 1 | term 2 | term 3 |

$$[RW]^*[IW]^*[IW]$$

term 4

then term 3 is replayed.

When we are dealing with the living recording medium of the brain, a natural question arises. How is the information recorded in the first place? Why is there a feeling or thought associated with the record?

To grasp this we need to look again at the parallel between sensation space and physical space. When a sensation is recorded what occurs is an imprint or shift of a ket in sensation space from a previous sensation, s, to a new sensation:

$$s + r \times w$$

where w is the information "bit" and r is a coupling constant that is adjustable according to the interaction between the observing record and the incoming bit, w. Usually we can regard the bits, w, as quantum units of information that are somehow well defined and separated. The sensations we have are not so well defined. In comparing two neighboring information bits, $w1$ and $w2$, with two neighboring sensations, $s1$ and $s2$, we usually have their difference:

$$\Delta s = s2 - s1$$

much smaller than the difference between two bits:

$$\Delta w = w2 - w1$$

after Δw is multiplied by the coupling constant, r. That is:

$$\Delta s << r \times \Delta w$$

This means that we are better able to resolve information than sensations. Or we need to involve many sensations in order to "see" any difference in external information. Small changes in sensations are unnoticed while small changes in objects are observable. This is just what is needed to make a good recording.

When we are very young our brains contain broad distributions of feeling possibilities with respect to sensations. If we label the initial or early child feeling ket:

$$|F[0]>$$

then the question bracket:

$$<s|F[0]>$$

which is the quamp that a feeling state, $F[0]$, has associated with it. a sensation, s, is broadly distributed over s.

The broadness of the distribution means that the young child is unable to associate his feeling, F, with any particular sensation, s. This appears as naiveté to "understanding" adults.

Using Law II of Chapter 8, we can resolve $<s|F[0]>$ into an infinite sum of sensation states as we did earlier in this chapter. This yields

$$<s|F[0]> = \text{SUM(over } s') <s|s'> <s'|F[0]>$$

In the following analysis this will be used to describe the resolution of sensations accompanying a wave of information.

Now we wish to look at how in time this distribution is altered by incoming information bits carried by quantum waves. We are looking at a young child's mind as he learns something new about his environment. The transformation of the "feeler," or engrams, of the brain as a result of recording the information interests us here. If the incoming wave moves through physical space x, then its quamp can be written:

$$<x|wi>$$

This means the quantum probability amplitude that the incoming wave bit wi is discovered at spatial location x in the brain or cortex is $<x|wi>$.

Suppose also that there is a drone or reference wave present:

$$<x|w0>$$

This exactly corresponds to the term $[RW]$ written above. This drone wave can be seen as a child rocks back and forth in his crib. Often autistic children rock in a similar manner. When we begin, the two abstract Hilbert space kets, $|wi>$ and $|s>$, are not interacting. This means that the state

$$|wi, s> = |wi>|s>$$

is a product of simple kets just as the probability of getting tails on a coin flip and rolling a die with the number 2 showing is a simple product of their individual probabilities.

Now suppose that a brain mechanism exists that is similar or common to those found in quantum-mechanical systems and that couples the sensation space to the external physical information wave space. After a single interaction, the sensing part of the ket "remembers" the wi. There is an operation that transforms the original simple ket, $|wi, s>$, into

$$|wi, s + rwi> = |wi>|s + rwi>$$

The child has now grasped a single bit of information as he grasps the finger of his father's outstretched hand. In other words, the external bit remains the same but the sensation state is shifted by the amount rwi.

This is the mechanism for making the brain hologram and it works as follows. Initially the sensing and space-occupying neural unit contains

$$<x|w0> <s|F[0]>$$

that is, it contains the drone reference wave alongside a similar drone feeling state. Then information from the outside world enters through the body senses. The child grabs the finger. This information, w, reaches the neural unit, where it adds to the original drone wave. Together they compose a sum of information bits represented as quantum waves in the brain space:

$$<x|w> = \text{SUM (over } wi \text{ states) } <x|wi> <wi|w>$$

Now the neural unit contains

$$[\text{SUM (over } wi) <x|wi> <wi|w>] \times <s|F[0]>$$

where each term of the sum multiplies the independent feeling state $s|F[0]>$. At this point, although the child has grabbed the finger, he doesn't "know" it yet. Next the interaction occurs between the brain medium and the information waves. Now the feeling state is no longer independent of the information. Each ket $|s'>$ in the resolution of $<s|F[0]>$ is altered. Going back to the sum over sensation states s' above, the neural unit now contains a double sum of coupled terms:

SUM (over wi) SUM (over s')
$$<x|wi> <wi|w><s'|s' + rwi> <s'|F[0]>$$

where the ket $|s' + rwi>$ now contains in the above a record of the information bit wi. Now the child knows it.

Since the sum over sensation states is so finely resolved, it is useful to group the terms in the sum under one banner. To do this let

$$|F[wi]> = \text{SUM (over } s') |s' + rwi> <s'|F[0]>$$

$|F[wi]>$ stands for a feeling ket containing the memory record wi. It is composed of a range of sensations, s', shifted by the small amount, rwi. Then the neural unit's contents are

$$\text{SUM (over } wi) <x|wi> <wi|w> <s|F[wi]>$$

Now the child's world has been unalterably changed. He has associated a range of sensations, Δs, with a single information bit, the grabbing of a finger. He can now distinguish, for example, mother's finger from father's. Or his own finger from that of either parent.

The term $<wi|w>$ is the probability amplitude that the wave w

contains the bit wi. The term $<x|wi>$ is the probability amplitude that the bit wi is marked at the place in the brain x. The term $<s|F[wi]>$ is the probability amplitude that the feeling state with the engram wi produces a sensation, s. This is the desired result.

The above sum shows how the location of an information bit, wi, located in the brain at x, is connected to emotional feelings, $F[wi]$, appearing as a bodily sensation, s.

This sum is nothing more than the quantum-mechanical representation of the sum of waves:

$$[RW] + [IW]$$

used to make the brain hologram. The only difference is that we have included the detail of what has occurred in the brain medium itself. It contains what might be called a "feeling tone" coupled to each possible information bit, wi. The sum of these waves are turned into physical representation when the complex-conjugate SUM (over wi) is multiplied with the above. Now the brain contains a permanent record of the probability waves distributed both through the whole brain space x and throughout the whole body sensation space s.

What originally was a pure feeling ket, $|F[0]>$, has now been transformed into an information-loaded (Jungians might use the term *polluted*) ket, $|F[wi]>$. No longer does the child feel the same way about the world. He has "taken a bite of the apple," wi, from the Tree of Knowledge in the primal world. He knows, and he no longer feels the same about it.

With each new wave of information, $<x|wj>$, coming into the neural unit, the above process repeats, only it starts where it left off. Instead of $|wj>|F[0]>$ — the new bit reaching the naive neural unit — it interacts with an old jaded one. We get

$$|wj>|F[wi]>$$

Now when a coupling occurs, the record contains

$$|wj>|F[wi + wj]>$$

After many such interactions the feeling state will contain a long sequence of memory bits, wi, wj, wk, ..., stored in whatever manner they were received, a kind of computer random access memory. Each time the feeling state is shifted by the amount rwj, corresponding to the added bit, wj, it is not the world that is altered, it is our brains that have gotten older.

To remember anything, these sequences are replayed every time a reference wave, $|w0>$, reverberates in our brains. Each neural unit occupies just enough physical space and sensation space to record a bit of the hologram. The sequence recorded in one neural unit need not contain the same sequence order as in a neighboring unit. However, the number of occurrences of each bit in each sequence will form a statistical distribution. If $w17$ occurs fifty times in unit 7, it will tend to occur an equal number of times in the other units. Just where in the chain $w17$ occurs depends on when $w17$ was received by a particular neural unit. In this manner the records tend to a statistical distribution of data and enable us to get on in the world. Even though Johnny learned to read the letter G just after B (Johnny is Greek), he has no problem with the alphabet.

It is thus through our feeling states in sensation space that we are conscious. It is through our sequences recorded quantum mechanically in the abstract Hilbert space of our minds that our memories are "loaded."

The Evolution of Consciousness — the Freedom of Accepted Insecurity

Each bit of memory mentioned above constitutes a single morpheme in the sensation space. Just as a wave of information is passed from a transmitter to a receiver, there is a corresponding movement of morphemes. The movement of morphemes in sensation is not actually a physical energy but is a "feeling wave" or tone. Our bodies do reflect this movement as physical energy. How we feel is a product of how these feeling waves move. Physical movement accompanies psychic movement.

R. Buckminster Fuller coined the word *synergy*. Perhaps synergy — the sympathetic action of two or more substances, organs, or organisms to achieve an effect of which either is incapable alone — is a better term. Morphemes move on the psychic level of reality just as matter moves on the physical level. A psycho/physical communication is synergistic. (It is interesting that, theologically speaking, synergy means the regeneration effected by the combination of human will and divine grace.)

Thus there are two parallel transformations: Morpheme sensation into feelings and matter into energy; only now we should look at the action taking place synergistically. Morphemes become feelings as matter becomes energy.

On the physical plane matter changes into energy only so long as there is a reason to do so. There must be an absorber as well as an emitter of that energy or else matter will not transform.

This idea is foreign to classical physics. An emitter will emit light, according to classical physics, regardless of its environment. But quantum physics requires the observer to have the photon strike an eye. The photon is only potentially present everywhere until there is an observer to see it as a flash of light on the retina. According to the *absorber theory of quantum physics*, the emitter will not emit unless it "senses" the presence of the absorber. This sensing takes place in the psychic plane. This sensing is the movement of "psychic light" from the emitter to the absorber. This "psychic light," or "psychic photon," is the result of transformation of morphemes into feelings. It is also difficult to pin down just what this carrier of feeling is. Mystics speak of an "inner light" and refer to ordinary light as the physical manifestation of the inner light. Perhaps the inner light and the outer light are the same. Although I hate to introduce more jargon into this discussion, perhaps we could call the inner light photon a *psychon.*

I have often been mystified by psychics who have told me that anything that happens has already happened. Perhaps that is what the "moving finger across the sky" concept of fate comes from. In the parallel worlds sense there is a real truth to this. Since there is no wave collapse caused by an outside consciousness, all events in all of the parallel worlds are predestined. But there is also a psychic sense of truth to this on its own grounds. The "finger in the sky" is the movement, on the psychic plane, of psychons from the emitter to the absorber. It is an agreement before the event or simultaneous with the sequence of events that take place on the physic plane. The physic emitter emits and the absorber absorbs the photon, provided that the "moving finger" on the psychic plane "wills" that it be so; or that there is a transaction of psychons between the two so that no movement will occur unless there is a mutual benefit to both, a synergistic effort of emitting and absorbing.

Nothing actually moves physically on the psychic plane because it is not physical. Space and time are confined to the physical plane; they play no role on the psychic plane except as ideals, pure thoughts. Yet the psychic plane can be grasped and, as I have shown, there is a mathematical way of understanding it. It won't hurt us to know this. It is our destiny on the psychic level to understand the psychic plane just as it is our destiny to understand the physical plane. Thousands of years ago we probably felt just as helpless about the physical world as we feel today about the psychic world. Many are terrified of psychic phenomena; but many feared fire in ancient days. Remember that fear is a psychic communication between the reptile and the man. When the man understands this, he is no longer ruled by the reptile and both the reptile and the man advance in consciousness. But if the man fears that he is a reptile and not capable of conquering his reptilian past, both the reptile and the man fall into the trap of rebirth.

This rebirth may be physical, but it need not be. Every time a man fears himself — his dark side — he is "rebirthing" as a reptile. That reptile exists not only as a reminder to us on the physical level but as an entity on another parallel universe, another physical space-time plane projection from the psychic plane. Man and reptile are one only on the psychic plane.

To escape from the fear traps, we need mathematics, and we need modern physics, especially quantum physics and relativity. By recognizing the mathematical structure of the psychic plane we will come to understand our psychic connection with the reptilian past, which is our present coexisting parallel universe.

To do this we need to understand the meaning of insecurity as the driving force of living evolution. We will always feel fear with something new, with each new thought. Fear is the basis for all emotions — i.e., fear and feeling are connected in the same manner as uncertainty and energy are connected in the physical world. Fear is the basic "energy" on the psychic plane and is the basis for feeling states on this plane. The origin of fear, like the origin of doubt and uncertainty, lies in the principle of uncertainty in quantum physics. Fear is as basic to life as love. Only mechanical devices have no fear, because they operate deterministically. As long as there is a tomorrow, there will be be fear.

By understanding fear we free ourselves of our desire for absolute security. Fear and the quest for security are actually the same things.

Strangely, once fear is accepted, it loses power. It is then able to be transformed, as any feeling is transformed, into sensation through thought. Fear and life go together as part and parcel of the evolution of species. Our present fixation with nuclear weapons, detente, and other "war college" concepts is the false attempt to externalize our fear. Even if there were no nuclear weapons, we all would die. Our fear of the planet's annihilation is purely an abstract concept based on the illusion that each of us will survive forever.

As I have tried to explain throughout the book, an understanding of the principles of quantum physics is necessary for conscious evolution. It is not just that the principles apply and therefore a passive understanding of them will explain evolution. This understanding is in itself an evolution of consciousness and makes the principles operational in life. The understanding becomes the experience of transformed and conscious living. The next step for us all will be a recognition that quantum physical principles, particularly the principles of uncertainty and complementarity, apply to our psychic lives just as deeply as they apply to our physical lives.

To evolve in a spirit of peace and blessed coexistence, human beings must understand quantum physics and its application to their minds and their consciousness. There may be no other way for this kind of conscious evolution. Through this understanding, which I have only begun to investigate, and which I feel is inevitable, human beings will realize a vast potential for a true spiritual evolution.

POSTSCRIPT

The Meaning of the "New Psychology"

The "new psychology," like the "new physics," alters radically our view of ourselves and our roles in the universe. The "new physics" opened the doors to this revelation. Just as the past can no longer be the cause and controller of the present in the physical world, it has also lost its hold on our mental worlds. The "new psychology" points not to the past but to the future.

The future is the only likely candidate for the cause of the force of evolution and the growth of human consciousness and intelligence. This neoteleological view of the role of the future means that time need no longer have its grip on us. If quantum mechanics is taken literally, then time is unobservable and, as I have pointed out in this book, parallel universes exist. These universes point to the remarkable creativity evident in the universe. All possibilities abound now, today. Each and every being is in this sense reborn each nanosecond. There is nothing back there in the past holding us to a fixed course. The past is our collected agreement to the meaning of the human condition in the present. This understanding not only alters the human psyche, it alters the artificial psyche as well.

Artificial intelligence, as it is designed today, is not true consciousness precisely because it contains no quantum uncertainty, no built-in possibilities for creativity arising out of indeterminism. In this book, perhaps for the first time, the real distinction between a human and a computer is made. The human is a quantum machine while the computer is a classical machine. The man makes both random ignorance errors and quantum indeterministic errors while the machine makes only one kind of error, errors of the first kind. The human's errors lead to thoughts and mistakes; the computer's errors lead only to mistakes. The human thinks; the computer follows a program. The new age in artificial intelligence will take place when a computer makes an error of the second kind.

John A. Wheeler, the renowned physicist and visionary, calls our new age of science Era III, the physics of meaning. The first age, Era I, gave us the geometrical paths of motion, as discovered by Galileo and Kepler, without providing any explanation of motion itself. The second age, Era II, gave us the laws of mechanical motion, as discovered by Newton, Maxwell, and Einstein, and even the chromodynamics of today's subnuclear matter, without providing any explanation of law. In the third age, Era III, we must seek the explanation of law itself.

In our new age the role of mind predominates. The tools of thinking, developed through the first two ages, are turned on themselves. Science, instead of appearing as the antispirit, now becomes the spirit's greatest champion. Positive thinking, altered consciousness, self-healing, and increased longevity become provinces of science, not mere wishful thinking. The whole universe becomes the home and nurture of mind.

If humans follow a computer program, it is a program of magnificent design. Its designer exists in the future. Each of us is a part of that program. Each of us follows a path. At each step of the path an infinite number of branches into the future are seen from the pathfinder's viewpoint. With billions of forks in the road, each pathfinder must choose the path he/she follows. The only guarantee each pilgrim has is that all paths lead to the same Designer. If this isn't God, then I don't know what God is.

EPILOGUE

The Quantum Physics Story of Creation . . . and the Final Dissolution

In the beginning was void. This void was constant.

LET THERE BE ACTION.

Let the symbol for that constant be A. A was nothing, a zero. The action was imagined. Call the symbol for that 0. Then doubt and uncertainty arose. Is this right? Is this correct? How to know? How to be? Whence epistomology? Ontology? A logic of knowing. A logic of being. A logic?

LET THERE BE DOUBT.

Let the symbol for that uncertainty be h — the fundamental error, fundamental doubt. Divide A by h. Divide A into billions of shards of doubt. Let doubt create further action in the universe.

ACTION. *Lights? Camera?*

But A remained zero. The void persisted.

LET THERE BE SOMETHING.

To create something from nothing, let there be increment and diminution, plus and minus. Let the plus be timelike and the minus spacelike.

$$A = \text{Timelike Phase} - \text{Spacelike Phase} = TP - SP$$

But now what? Make the timelike phase TP a product.

LET THERE BE TIME.

$$TP = (\text{time-frequency}) \times \text{time} = tf \times T$$

Time appears jointly with frequency. Frequency means repetition. Repetition means memory. Let there be memory. And memory brings forth time. Time-frequency means energy.

But there was no time. There was no memory. Not yet.

LET THERE BE SPACE.

Make the spacelike phase, SP, a product.

$$SP = (\text{space-frequency}) \times \text{space} = sf \times S$$

Space appears jointly with space frequency. Frequency means repetition. Repetition means extension. Let there be extension. And extension brings forth space. Space-frequency means movement. Movement means momentum. But there was as yet no space. There was as yet no movement.

Space appears jointly with space frequency. Time appears cojointly with time-frequency. Divide time-frequency by space-frequency. Call the ratio *c*.

$$c = \text{time-frequency/space-frequency} = tf/sf$$

Let there be movement in this ratio.

LET THERE BE LIGHT.

ACTION. LIGHTS. *Camera?*

And there *was* light.

The ratio *c* became the speed of light. But there was still nothing. Light is action, is zero.

$$A = tf \times T - sf \times S = 0$$

But there is light. There is space? There is time? Let *A* not be zero.

The mind then moved. It stirred. A feeling of great violence appeared in its "gut" of consciousness. Make the negation of all that was made. Make not-*A*. Make the anti-*A*.

And then there was the big bang. Matter and antimatter appeared as knots and whirls in the mind, exploding in parallel space-times from an impossible gathering in imagination. *A*, the absolute phase of the void of consciousness, appeared and saw itself in an infinite hall of mirrors. It said:

LET THERE BE MATTER.

And *A* became what it was not. It became not nothing. Not void. It became something.

$$0 = A + (\text{anti-}A) = A + \sim A$$

And now *A* was something. But it was not matter yet. It was matter-phase. *A* was the phase of matter. And it was divided by *h*. It was still uncertain. There was still doubt.

$$A = mc^2 \times T/h$$

And anti-*A* was something. It was the antimatter-phase.

And matter was energy (when multiplied by the speed of light squared). And energy was frequency (when divided by *h*). And that meant time.

$$A = mc^2 \times T/h = E \times T/h = tf \times T$$

And there was light.

ACTION. LIGHTS. CAMERA.

Now there were "viewpoints." These viewpoints were all seeing A differently. Some saw it as matter everywhere potentially able to be. Others saw it as matter moving faster than light potentially able to be. But there were no viewers. Not yet.

$$A = (mc^2 T)/h = (Et - MS)/h$$

For one it was matter "sitting." For another it was matter moving with energy E and momentum M. Yet it was always potentially sitting for one and potentially moving faster than light for the other; for the time-frequency E divided by the space-frequency M was bigger than c.

And that was relativity. It was a question of phase that was related to action. And the constancy of that phase was always the same, but the frequencies and the time and spatial intervals were different. And there appeared no order or reason for anything.

And out of this phase of matter a gigantic hologram was constructed and interwoven with itself. The warp and woof were made from frequencies that, when viewed from one perspective, appeared as traveling waves, each in a different universe weaving into each other in an infinite-dimensional manner.

A group of waves, appearing to travel together for a slight instant, managed to slow things down. The group that really didn't exist yet was composed of waves in layers, each layer in a different universe. That momentary gathering of ripples traveled at the group velocity determined by the difference in time-frequencies divided by the difference in space-frequencies in the different layers.

$$\text{Group Velocity} = GV = \Delta ts/\Delta sf = \Delta E/\Delta M$$

And GV was less than c. And then consciousness noticed that this preserved order in the universes. The group that managed to appear as a single bump of waves (provided "one" was able to extract "oneself" from the hologram and see all the universes at once) traveled together from one point of view (if the viewer existed). That is, it moved in such a way that there was no point of view (if any other viewer existed) to contradict that movement. This was not true for any of the simple single-universe waves from which the "group" was constructed.

When seen as traveling waves, the simple single-universe waves would appear to move forward from one point of view and backward from another. This was because they were phase waves moving faster than light.

And consciousness saw an idea in this.

LET THERE BE PAST AND FUTURE.

Past shall be defined by these funny little bumpy groups. They appear to come from somewhere. Call that "past." They appear to be going somewhere. Call that "future." And consciousness imagined itself as made of these groups. It called these groups "particles."

And it learned to focus on these particles. And then it occurred to consciousness how to make it come alive. A gigantic entertainment center filled with drama and pathos, love and chaos. It became its dream. And with that thought

CONSCIOUSNESS AND THE BIG BANG SEPARATED.

And there was darkness, for the matter, being part of the dreamer, knew from where it came out of light and out of void. It knew it was light. It was trapped light. And it cried out for freedom. It bubbled and gurgled in impossible gyrations. And there was spin. Each bubbling sent spontaneous needles of light seeking out in the darkness. Each light photon returned. But then one day the light did not return. The electron had discovered another electron. And then there was quantum electrodynamics. And there was light interacting with matter. And there was knowing. And there was being. Consciousness had created the means of its knowing itself.

Actually, consciousness hadn't separated totally from the hologram. It had just gotten some of itself ensnared in the warp and woof of it all. It was, so to speak, in two places at the same time — outside of all of the space-times that were the hologram, and in tiny pockets of "bundled" groups of waves made up of layers of the hologram folded over on each other like a fine-mesh lace curtain blown gently across itself by the wind.

Those "groupie-waves" enfolded and unenfolded themselves, reappearing at different places in the fabric of the hologram.

But not really. The whole thing just *was.* It was consciousness itself that made the hologram move. There was no time for all that movement to take place. Only consciousness moved. It was all in the viewing. That was the trick. It was all a superoptical illusion. Light was consciousness. Consciousness was entrapped light. Consciousness was outside of the entrapment too. From the outside perspective a scan of the hologram showed moving events. Groups of waves appeared only when the viewer superposed parallel layers together. It was all in the perspective. These wave groups showed themselves as things having real solid dimensions, all moving with speeds — on the average — slower than light-speed, slower than the viewer.

And that occurred in an orderly way, following what appeared to be laws of causality based on the past. But the waves in each universe layer moved their phases faster than light, because they were everywhere at the same time. Not time. No time. Not space. No space. Just is.

And consciousness thought: Time to get free. Got to make the hologram realize that it is me trapped. Got to find the way. No good to use past. The groups are already following patterns based on where they were. Can't break those patterns because they are addicted to the repeat cycle. They enjoy that "old trapped feeling." Start from the future. Get enough groups to give up their "groupie" tendencies and evolve. To do this "I" must be more involved.

LET MATTER BE CONSCIOUS.

Let the groups entrap me more. Become the world hologram.

And at the end of time — which was only the edge of the infinite-dimensional hologram — consciousness entered the hologram completely. Consciousness became its own illusion. The hologram was seen to be its own evolution.

And outside of time the hologram had changed. The future beckoned anticausally to the wave groups caught in the myths of their own logos. The groups hoped and dreamed.

And all the universes in the whole hologram awoke. Consciousness became matter. And the hologram vanished in a dream. Consciousness awoke and . . .

BIBLIOGRAPHY

A & Z (Appignanesi, Richard, and Zarate, Oscar). *Freud for Beginners.* New York: Pantheon Books, 1979.

Adams, Douglas. *The Restaurant at the End of the Universe.* New York: Pocket Books, 1980.

Armstrong, Herbert W. "Why Must Man Suffer?" *The Plain Truth* 9 (Oct. 1983): p. 19.

Aspect, Alain, Dalibard, Jean, and Roger, Gérard. "Experimental Test of Bell's Inequalities Using Time-Varying Analyzers." *Physical Review Letters* 25 (Dec. 1982): 1804.

Bass, L. "A Quantum Mechanical Mind-Body Interaction." *Foundations of Physics* 1 (1975): 159.

Bateson, Gregory. *Mind and Nature: A Necessary Unity.* New York: E. P. Dutton, 1979.

Beahrs, John. *Unity and Multiplicity: Multilevel Consciousness of Self in Hypnosis, Psychiatric Disorder and Mental Health.* New York: Bruner/Mazel, 1983.

Beckenstein, J. "Black Holes and Entropy." *Physical Review Digest,* No. 8 (1973): 2333.

Bell, J. S. "On the Einstein Podolsky Rosen Paradox." *Physics* 1 No. 1 (1964): p. 195.

Benioff, Paul. "Quantum Mechanical Models of Turing Machines That Dissipate No Energy." *Physical Review Letters* 23 (June 1982): 1581.

Bennett, C. H. "Logical Reversibility of Computation." *IBM Journal of Research and Development* 17 (Nov. 1973): 525.

Bohr, Niels. *Atomic Physics and Human Knowledge.* New York: Wiley, 1958.

Brillouin, Leon. *Science and Information Theory.* New York: Academic Press, 1962.

Broyles, A. A. "Derivation of the Probability Rule." *Physical Review Digest* 12 (June 1982): 3230.

Buber, Martin. *I and Thou.* Translated by Walter Kaufmann. New York: Charles Scribner's, 1970.

Buksbazen, John Daishin. *To Forget the Self.* Vol. III Zen Writings Series. Los Angeles: Zen Center of Los Angeles, no date.

Calvin, Melvin. *Chemical Evolution: Molecular Evolution Towards the Origin of Living Systems on the Earth and Elsewhere.* New York: Oxford University Press, 1969.

Chaitin, Gregory J. "Information-Theoretic Computational Complexity." *IEEE Transactions on Information Theory* 1 (Jan. 1974): 10.

Clark, Ronald W. *Einstein: The Life and Times.* New York: World Publishing, 1971.

Coultier, Neal S. "Software Science and Cognitive Psychology." *IEEE Transactions on Software Engineering* 2 (March 1983): 166.

Cramer, John G. "Generalized Absorber Theory and the Einstein-Podolsky-Rosen Paradox." *Physical Review Digest* 22 (1980): 362.

Da Free John. *The Transcendence of Ego and Egoic Society.* Clearlake, Cal.: The Johannine Daist Communion, 1982.

— —. *A Call for the Radical Reformation of Christianity.* Clearlake, Cal.: The Johannine Daist Communion, 1982.

— —. *The Transmission of Doubt.* Clearlake, Cal.: The Dawn Horse Press, 1983.

— —. *Easy Death.* Clearlake, Cal.: The Dawn Horse Press, 1983.

David-Neel, Alexandra. *Buddhism: Its Doctrines and Its Methods.* New York: Avon Books, 1979.

David-Neel, Alexandra, and Yongden, Lama. *The Secret Oral Teachings in Tibetan Buddhist Sects.* San Francisco: City Lights Books, 1967.

Davis, Philip J., and Hersh, Reuben. *The Mathematical Experience.* Boston: Houghton Mifflin Co., 1982.

Dennett, Daniel C. *Brainstorms: Philosophical Essays on Mind and Psychology.* Cambridge, Mass.: The MIT Press, 1981.

DeWitt, B., and Graham, N., eds. *The Many-Worlds Interpretation of Quantum Mechanics.* Princeton: Princeton University Press, 1973.

Dirac, P. A. M. *The Principles of Quantum Mechanics.* London: Oxford University Press, 1958.

Ditfurth, Hoimar von. *The Origins of Life: Evolution as Creation.* Translated by Peter Heinegg. San Francisco: Harper & Row, 1982.

Duncan, Ronald, and Weston-Smith, Miranda, eds. *The Encyclopaedia of Ignorance: Everything You Ever Wanted to Know About the Unknown.* Elmsford, N.Y.: Pergamon Press, 1977.

Eccles, John C. "The Physiology of Imagination." In *Altered States of Awareness.* Readings from *Scientific American.* San Francisco: W. H. Freeman & Co., 1971.

Edelman, Gerald M., and Mountcastle, Vernon B. *The Mindful Brain: Cortical Organization and the Group-Selective Theory of Higher Brain Function.* Cambridge, Mass.: The MIT Press, 1982.

Einstein, A., Lorentz, H. A., Weyl, H., and Minkowski, H. *The Principle of Relativity.* New York: Dover Publications, 1923.

Einstein, Albert, Podolsky, Boris, and Rosen, Nathan. "Can Quantum-Mechanical Description of Physical Reality Be Considered Complete? *Physical Review* 47 (1935): 777.

Eiseley, Loren. *The Man Who Saw Through Time.* New York: Charles Scribner's, 1973.

Eisenberg, D., and Kauzmann, W. *The Structure and Properties of Water.* New York: Oxford University Press, 1969.

Elvee, Richard G., ed. *Mind in Nature.* Nobel Conference XVII. San Francisco: Harper & Row, 1982.

Fadiman, James, and Frager, Robert. *Personality and Personal Growth.* New York: Harper & Row, 1976.

Feynman, Richard P. *Quantum Electrodynamics.* New York: W. A. Benjamin, 1961.

Feynman, Richard P., Leighton, Robert B., and Sands, Matthew. *The Feynman Lectures on Physics: Quantum Mechanics.* Reading, Mass.: Addison-Wesley Publishing, 1965.

Fincher, Jack. *Human Intelligence.* New York: G. P. Putnam's, 1976.

Fordman, Frieda. *An Introduction to Jung's Psychology.* New York: Penguin Books, 1953, 1978.

Francon, Maurice. *Holography.* Expanded and revised from the French Edition. New York: Academic Press, 1974.

Franson, J. D. "Extension of the Einstein-Podolsky-Rosen Paradox and Bell's Theorem." *Physical Review* 4 (1982): 787.

Freedman, David, Pisani, Robert, and Purves, Roger. *Statistics.* New York: W. W. Norton & Co., 1978.

Freud, Sigmund. *An Outline of Psycho-Analysis.* New York: W. W. Norton & Co., 1949.

— —. *The Ego and the Id.* New York: W. W. Norton & Co., 1960.

— —. *The Future of an Illusion.* New York: W. W. Norton & Co., 1961.

— —. *Civilization and Its Discontents.* New York: W. W. Norton & Co., 1961.

— —. *Group Psychology and the Analysis of the Ego.* New York: Boni and Liveright, 1924.

— —. *Psychopathology of Everyday Life.* New York: Macmillan, date unknown.

— —. *The Interpretation of Dreams.* New York: Avon Books, 1965.

Fromm, Erich. *Greatness and Limitations of Freud's Thought.* New York: Harper & Row, 1980.

— —. *The Sane Society.* New York: Holt, Rinehart & Winston, 1955.

Furst, Charles. *Origins of the Mind: Mind-Brain Connections.* Englewood Cliffs, N.J.: Prentice-Hall, 1979.

Godwin, Joscelyn. *Mystery Religions in the Ancient World.* San Francisco: Harper & Row, 1981.

Govinda, Lama Anagarika. *Foundations of Tibetan Mysticism.* London: Rider & Co., 1960.

Gray, William, and La Violette, Paul. Interviewed in *The Brain Mind Bulletin* 6 (March 1982): 1 and 7 (March 1982): 1.

Gregory, Richard L. *Mind In Science: A History of Explanations in Psychology and Physics.* Cambridge: Cambridge University Press, 1981.

Gooch, Stan. "Right Brain, Left Brain." *New Scientist* 11 (Sept. 1980): 790.

Haich, Elisabeth. *Sexual Energy and Yoga.* New York: ASI Publishers, 1972.

Hall, Calvin S. *The Meaning of Dreams.* New York: McGraw-Hill, 1966.

Hampden-Turner, Charles. *Maps of the Mind.* New York: Macmillan, 1981.

Harth, Erich. *Windows on the Mind: Reflections on the Physical Basis of Consciousness.* New York: William Morrow & Co., 1982.

Highwater, Jamake. *The Primal Mind: Vision and Reality in Indian America.* New York: Harper & Row, 1981.

Hofstadter, Douglas R. *Gödel, Escher, Bach: An Eternal Golden Braid.* New York: Basic Books, 1979.

Hoyle, Fred. "The Universe: Past and Present Reflections." *Astrophysics and Relativity,* Preprint Series No. 70. Cardiff, Wales: Department of Applied Mathematics and Astronomy, University College, 1981.

Hubel, David H., and Wiesel, Torsten N. "Brain Mechanisms of Vision." *The Brain.* San Francisco: W. H. Freeman & Co., 1979.

Jahn, Robert G. "The Persistent Paradox of Psychic Phenomena: An Engineering Perspective." *Proceedings of the Institute of Electrical and Electronics Engineers (IEEE)* 2 (Feb. 1982): 136.

Jammer, Max. *The Philosophy of Quantum Mechanics.* New York: John Wiley, 1974.

——. *The Conceptual Development of Quantum Mechanics.* New York: McGraw-Hill, 1966.

Jampolsky, Gerald G. *Love Is Letting Go of Fear.* Millbrae, Cal.: Celestial Arts, 1979.

Jones, Roger S. *Physics as Metaphor.* Minneapolis, Minn.: University of Minnesota Press, 1982.

Jung, C. G. *The Undiscovered Self.* Boston: Little, Brown & Co., 1957.

——. *Analytical Psychology: Its Theory and Practice.* New York: Vintage Books, 1968.

——. *Man and His Symbols.* New York: Dell Publishing, 1968.

——. "Synchronicity: An Acausal Connecting Principle." In *The Interpretation of Nature and the Psyche.* W. Pauli and C. G. Jung, eds. Princeton, N.J.: Pantheon, 1955.

——. *On the Nature of the Psyche.* Princeton, N.J.: Princeton University Press, 1960.

——. *Answer to Job.* Princeton, N.J.: Princeton University Press, 1958.

Kandel, Eric R., and Schwartz, James H. "Molecular Biology of Learning: Modulation of Transmitter Release." *Science* 29 (Oct. 1982): 433.

Kaempffer, F. A. *Concepts in Quantum Mechanics.* New York: Academic Press, 1965.

——. *The Elements of Physics: A New Approach.* Waltham, Mass.: Blaisdell Publishing, 1967.

Kaufmann, Walter. *Religions in Four Dimensions.* New York: Reader's Digest Press, 1976.

Kline, Morris. *Mathematics: The Loss of Certainty.* New York: Oxford University Press, 1980.

Klopf, A. Harry. *The Hedonistic Neuron: A Theory of Memory, Learning, and Intelligence.* Washington, D.C.: Hemisphere Publishing, 1982.

Laing, R. D. *The Voice of Experience.* New York: Pantheon Books, 1982.

Mandelbrot, Benoit B. *Fractals: Form, Chance and Dimension.* San Francisco: W. H. Freeman & Co., 1977.

McCorduck, Pamela. *Machines Who Think.* San Francisco: W. H. Freeman & Co., 1979.

Miller, Jonathan. *States of Mind.* New York: Pantheon Books, 1983.

Morris, W., ed. *The American Heritage Dictionary of the English Language.* Boston: American Heritage Publishing and Houghton Mifflin, 1969.

Mumford, Lewis. *Technics and Human Development: The Myth of the Machine.* Vol. One. New York: Harcourt Brace Jovanovich, 1967.

Munitz, Milton K. *Space, Time and Creation: Philosophic Aspects of Scientific Cosmology.* Glencoe, Ill.: The Free Press, 1957. (Also see edition published in New York: Dover Publications, 1981)

Narliker, Jayant. *The Structure of the Universe.* Oxford, England: Oxford University Press, 1977.

Nelson, Benjamin, ed. *Freud and the 20th Century.* Cleveland: World Publishing, 1957.

Ornstein, Robert E. *The Psychology of Consciousness.* San Francisco: W. H. Freeman & Co., 1972.

— —. ed. *The Nature of Human Consciousness: A Book of Readings.* New York: Viking Press, 1973.

Ouspensky, P. D. *In Search of the Miraculous.* New York: Harcourt, Brace & World, 1949.

— —. *The Fourth Way.* New York: Random House, 1957.

— —. *Tertium Organum: A Key to the Enigmas of the World.* New York: Vintage Books, 1970.

Pelletier, Kenneth R. *Toward a Science of Consciousness.* New York: Dell Publishing, 1978.

Penfield, Wilder. *The Mystery of the Mind.* Princeton, N.J.: Princeton University Press, 1975.

Piaget, Jean. *The Child's Conception of Time.* New York: Ballantine Books, 1971.

Pribram, Karl H. *Languages of the Brain: Experimental Paradoxes and Principles in Neuropsychology.* Monterey, Cal.: Brooks/Cole Publishing, 1977.

Prigogine, Ilya. *From Being to Becoming: Time and Complexity in the Physical Sciences.* San Francisco: W. H. Freeman & Co., 1980.

Progoff, Ira. *Jung, Synchronicity, and Human Destiny: Noncausal Dimensions of Human Experience.* New York: Dell Publishing, 1973.

Ramon-Moliner, E. "Space-Time and Neural Events." *Brain Theory Newsletter* 3 (March 1977): 52.

— —. Personal correspondence, dated October 19, 1982.

Restak, Richard M. *The Brain: The Last Frontier.* New York: Doubleday, 1979.

— —. "Is Free Will a Fraud?" *Science Digest* (Oct. 1983): 52.

Rosenblatt, Allen D., and Thickstun, James T. "Modern Psychoanalytic Concepts in General Psychology. Part One: Concepts and Principles. Part Two: Motivation." *Psychological Issues* 2/3, Monograph Series 42/43. New York: International Universities Press, 1977.

Rosenfeld, Albert. *Mind and Supermind.* New York: Holt, Rinehart and Winston, 1977.

Ross, Nancy Wilson. *Buddhism: A Way of Life and Thought.* New York: Vintage Books, 1981.

Rothenberg, Albert. *The Emerging Goddess: The Creative Process in Art, Science and Other Fields.* Chicago: University of Chicago Press, 1979.

Sandburg, Carl. *Harvest Poems: 1910-1960.* New York: Harcourt Brace Jovanovich, 1960.

Schatzman, Morton. "Evocations of Unreality." *New Scientist* 25 (Sept. 1980): 935.

— —. *The Story of Ruth.* London: Duckworth, 1980.

Shannon, C. E. "A Mathematical Theory of Communication." *The Bell System Technical Journal* 3 (July 1948): 379.

Shannon, Claude E., and Weaver, Warren. *The Mathematical Theory of Communication.* Urbana, Ill.: University of Illinois Press, 1963.

Skinner, B. F. *Science and Human Behavior.* New York: The Free Press, Macmillan, 1953.

Stevens, Peter S. *Patterns in Nature.* Boston: Little, Brown & Co., 1974.

Suares, Carlo. *The Cipher of Genesis: The Original Code of the Qabala as Applied to the Scriptures.* Berkeley, Cal.: Shambhala, 1970.

— —. *The Resurrection of the Word.* Berkeley, Cal.: Shambhala, 1975.

— . *The Song of Songs: The Canonical Song of Solomon Deciphered according to the Original Code of the Qabala.* Berkeley, Cal.: Shambhala, 1975.

— —. *Les Spectrogrammes de l'alphabet hebraique.* Geneva, Switz.: Mont-Blanc, 1973.

— —. *The Second Coming of Reb Yhshwh, The Rabbi Called Jesus Christ.* Unpublished English translation of *Memoire sur le Retour du Rabbi qu'on appelle Jesus.* Paris: Robert Lafont, 1975.

Taylor, Gordon Rattray. *The Natural History of the Mind.* New York: E. P. Dutton, 1979.

Tibbetts, Paul, ed. *Perception: Selected Readings in Science and Phenomenology.* New York: Quadrangle/The New York Times Book Co., 1969.

Thompson, William Irwin. *Passages About Earth: An Exploration of the New Planetary Culture.* New York: Harper & Row, 1973.

Thomsen, Dietrick E. "A Knowing Universe Seeking to be Known." *Science News* 19 (Feb. 1983): 124.

Tulku, Tarthang. "The Life Story of Shakyamuni Buddha." *Crystal Mirror.* Vol. III. Emeryville, Cal.: Dharma Publishing, 1974.

von Franz, Maria-Louise. *Number and Time: Reflections Leading to a Unification of Depth Psychology and Physics.* Evanston, Ill.: Northwestern University Press, 1974.

— —. *Time: Rhythm and Response.* New York: Thames and Hudson, 1978.

Wheeler, John Archibald. "Delayed Choice Experiments and the Bohr-Einstein Dialog." Paper presented at the American Philosophical Society and Royal Society Joint Meeting, London, 1980.

Wiener, Norbert. *God & Golem, Inc.: A Comment on Certain Points where Cybernetics Impinges on Religion.* Cambridge, Mass.: The MIT Press, 1964.

Wilber, Ken, ed. *The Holographic Paradigm and Other Paradoxes.* Boulder, Col.: Shambhala, 1982.

Wolf, Fred Alan. *Taking the Quantum Leap: The New Physics for Nonscientists.* San Francisco: Harper & Row, 1981.

Woo, C. H. "Consciousness and Quantum Interference—An Experimental Approach." *Foundations of Physics* 11/12 (1981): 933.

Yogananda, Paramahansa. *Science of Religion.* Los Angeles: Self-Realization Fellowship Press, 1972.

INDEX

339